On a clear night, the vastness and beauty of the star-filled sky is awe-inspiring. In *Stargazing: Astronomy Without a Telescope* Patrick Moore will tell you all you need to know about the universe visible to the naked eye. With the aid of charts and illustrations he explains how to 'read' the stars, to know which constellations lie overhead, their trajectory throughout the seasons, and the legends ascribed to them. In a month by month guide he describes, using detailed star maps, the night skies of both the northern and southern hemispheres. He also takes a look at the planets, the Sun and the Moon and their eclipses, comets and meteors, as well as aurorae and other celestial phenomena – all in accessible scientific detail. Astronomy is for everyone, and even with just the naked eye, it can become a fascinating and rewarding hobby for life.

PATRICK MOORE, lunar cartographer and host of the monthly television show *The Sky at Night* since 1957, is a distinguished and prolific author of more than 70 astronomy books. He has served as President of the British Astronomical Association. His contributions to the public understanding of astronomy have been marked by special awards from the Astronomical Society of the Pacific, the Italian Astronomical Society and the British Astronomical Association. Patrick Moore is a great enthusiast, always encouraging anyone with an interest in the night sky to go out and observe.

Acknowledgements

My thanks are due to Dr Simon Mitton and Dr Alice Houston, of Cambridge University Press, for their help and encouragement. I am also grateful to Dr Lindsay Nightingale for very helpful comments.

PATRICK MOORE
Selsey, May 2000

Stargazing

Astronomy Without a Telescope

SECOND EDITION

Patrick Moore

PUBLISHED BY THE PRESS SYNDICATE OF THE UNIVERSITY OF CAMBRIDGE
The Pitt Building, Trumpington Street, Cambridge, United Kingdom

CAMBRIDGE UNIVERSITY PRESS
The Edinburgh Building, Cambridge CB2 2RU, UK
40 West 20th Street, New York, NY 10011–4211, USA
10 Stamford Road, Oakleigh, VIC 3166, Australia
Ruiz de Alarcón 13, 28014 Madrid, Spain
Dock House, The Waterfront, Cape Town 8001, South Africa

http://www.cambridge.org

First published 1985 by Aurum Press
Second edition published by Cambridge University Press 2001

Printed in the United Kingdom at the University Press, Cambridge

Typeface Palatino 10/12 pt *System* QuarkXPress® [DS]

A catalogue record for this book is available from the British Library

Library of Congress Cataloguing in Publication data

Moore, Patrick.
Stargazing: astronomy without a telescope/Patrick Moore.– 2nd ed.
p. cm.
Includes index.
ISBN 0 521 79052 2
1. Astronomy–Observers' manuals. I Title.
QB63.M66 2000
523–dc21 00-037884

ISBN 0 521 79052 2 hardback
ISBN 0 521 79445 5 paperback

Contents

1

Astronomy and astronomers

What is the conventional idea of an astronomer? The old picture was that of an elderly man with a long white beard, sitting in a lonely observatory throughout the night 'watching the stars'. This picture was never true, and today it could not be less accurate. Your professional astronomer is much more likely to be a youngish, energetic person who spends very little time at the eye-end of a telescope. Instead, all his work is carried out by electronic methods, and he can remain in the comfort of a warm control-room when the telescope is being used. Nowadays it is not even necessary for the astronomer to be in an observatory at all. A large telescope in, for instance, the Mauna Kea observatory on top of a volcano in Hawaii can be operated by an astronomer in his office in London or Boston.

Moreover, professional astronomy is theoretical, and every hour spent in obtaining information from the sky means a great many dozens of hours of desk work. Strange though it may seem, I know a number of eminent professionals who would be quite unable to go out on a clear night and identify the various star-groups!

Yet there is another side to astronomy. The amateur with no professional qualifications whatsoever can still make valuable contributions, simply because he looks at the sky and knows it well. Some amateurs make a hobby out of hunting for comets and exploding stars or novae, while others (such as myself) are much more at home when carrying out telescopic observations of the Moon and planets. Despite all the new techniques, and despite the spacecraft which have caused such a revolution in outlook over the past few decades, there is still a great deal which we do not know.

Remember, too, that those people who have no wish to become 'serious astronomers', either professional or amateur, can enjoy themselves immensely by making the acquaintance of the stars. I began to take a real interest at the age of six (when, admittedly, the situation was very different from that of today), and my first step was to go outdoors with a star map, wait until the sky was dark and clear, and then learn my way around. It did not take long, and in fact it is not in the least difficult.

It is for these casual enthusiasts that I am writing now. Let us assume that you know absolutely nothing about astronomy, and have no wish to spend money on a telescope or even a pair of binoculars. The naked eye alone is quite adequate; so we will begin at the very beginning, with the fundamental facts. Anyone who knows them already has my full permission to skip the next few pages and proceed straight to Chapter 2!

First, there is a common error to be corrected. Some people still confuse astronomy with astrology. Actually, the two are as different as the proverbial chalk and cheese. Astronomy, the study of the sky and everything in it, is an exact science. Astrology is mediaeval superstition. Astrologers claim to link the positions of the Sun, Moon and planets with the characters and destinies of men and women born at different times of the year and in different places, so that a baby born on, say, 1 March in London will have a 'horoscope' quite different from that of a baby who first saw the light of day in Bognor Regis on 10 April. The only polite way to describe astrology in a single word is: 'Rubbish'. It has no basis of fact, and is strictly for the credulous only. The best that can be said of it is that it is fairly harmless when treated as a parlour game.

The Earth, our home in space, is a planet: a globe 7926 miles in diameter, moving round the Sun at a mean distance of 93 million miles. The Sun itself is a star, no larger, brighter or more luminous than many of the stars you can see on any clear night. The only reason why the Sun appears so splendid is that on the astronomical scale it is close to us, and this brings us on to the first difficulty: the problem of visualizing tremendous distances and vast spans of time. Frankly, nobody can really appreciate even one million miles, to say nothing of 93 million, but we know that the figures given are correct, and we simply have to accept them. As I have said often enough, astronomers cannot appreciate vast distances and time-scales any better than anyone else; the only difference is that they don't make the mistake of trying.

A very rough analogy may help. Suppose that we set out to fly to the Sun, moving at a steady 100 m.p.h. and never stopping. In one day we will cover approximately 2400 miles – but it will take us more than a hundred years to reach the Sun.

The Earth is not the only planet. Moving round the Sun there are eight others, of which the best known are Mercury, Venus, Mars, Jupiter and Saturn. They have no light of their own; they shine only because they reflect the light of the Sun, in the same way that a tennis ball will do if you shine a torch on it in a darkened room (though in fact most of the planets are much less reflective than tennis balls). Though the planets look like stars, and some of them are brilliant, they are our near neighbours, and are very junior members of the universe as a whole. Some of them have satellites moving round them; we have only one satellite (the Moon), but Saturn has as many as eighteen. The Sun's family, or Solar System, is completed by various objects of lesser importance such as comets and meteors, which I will describe later.

The Solar System itself is a very insignificant unit. The stars lie at immense distances, measured in millions of millions of miles. Represent the distance between the Sun and the Earth by one inch, and the nearest star will be over four miles away. This is why the stars appear as tiny points of light. It is also why they seem to keep virtually the same positions in the sky relative to each other – a fact which is of such fundamental importance that I must pause to say rather more about it.

Consider two flying objects: a sparrow at tree-top height, and a jet aircraft many miles up. The sparrow will move quickly against its background, while the jet will appear to crawl, but in reality the jet is much the faster of the two (at least, I have no record of any sparrow which can break through the sound barrier). The jet seems to move much more slowly merely because it is so much further away. The rule is: 'The further, the slower', and the stars are so remote that relative to each other they do not move detectably, as seen by the naked-eye observer, even over periods of many lifetimes. The patterns or constellations which we see today are to all intents and purposes the same as those which must have been seen by George Washington, Alfred the Great, Julius Caesar and the builders of the Pyramids. True, the stars do have tiny individual or 'proper' motions, because they are moving through space in all sorts of directions at all sorts of speeds, but for the moment we may regard the stars as 'fixed'.

Not so the Sun, Moon and planets, which are much closer to us, and wander slowly around the sky from one constellation to another. They do, however, keep within certain definite limits, and this was how they were identified by the ancient astronomers; the very word planet really means 'wanderer'.

Obviously it is impossible to plot the planets on permanent star-maps. This is also true of the Moon, which is the nearest natural body in the sky and is the only one which revolves round the Earth. (To be accurate, the Earth and Moon are revolving together round the centre of gravity of the Earth–Moon system, but for the moment we are dealing with fundamentals.) The Moon has a diameter of only 2160 miles, and its mean distance from us is 239,000 miles. If you want to make a reasonable though rough model, take a tennis ball to represent the Earth, wrap a piece of string around it ten times, and then unravel the string and put a table-tennis ball on the far end to represent the Moon. A simple calculation shows that the Sun is about 400 times as far away from the Earth as the Moon.

Like the planets, the Moon has no light of its own. It depends upon sunlight, and clearly only half of it can be illuminated at any one moment. This is why the Moon shows its regular phases, or apparent changes of shape from new to full. The mean interval between one full moon and the next is 29½ days, and when nearly full the Moon is brilliant enough to drown all but the brighter stars, even though it would take more than half a million full moons to equal the brightness of the Sun.

How many stars can you see with the naked eye on a dark night? When I put this question recently to a group of newcomers to astronomy, the answers ranged between 'ten thousand' and 'millions'. Rather surprisingly, nobody can ever see more than about 2500 stars at any one time; appearances can be deceptive. Of course, optical aid gives a prompt increase. The total number of stars in our star system or Galaxy is of the order of 100,000 million, and even this is only a beginning. Beyond our Galaxy there are others, each with its quota of stars.

Come back to the question of distance. We have seen that the stars are very remote, and it becomes awkward to give their distances in miles or kilometres, just as it would be clumsy to give the distance between London and New York in inches or centimetres. Fortunately there is a better unit available. Light does not travel instantaneously; it flashes along at 186,000 miles per second, so that in a year it can cover 5,880,000,000,000 or rather less than six million million miles. This is the unit known as the light-year – a unit not of time, please note, but of distance. If you want to convert light-years to miles, multiply by six million million and you will not be so very far out. The nearest star beyond the Sun is 4.2 light-years away, which works out at approximately 24 million million miles. The Galaxy measures 100,000 light-years from one side to the other; the distances of external galaxies range out to thousands of millions of light-years.

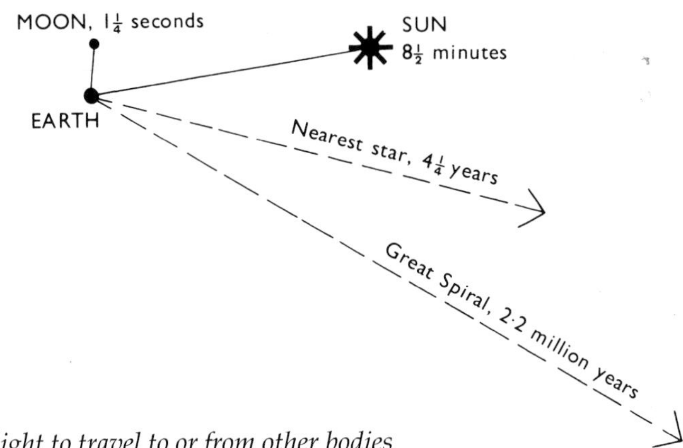

Time taken by light to travel to or from other bodies

There is an interesting result of this. Suppose that we look at a galaxy two million light-years away, as we can do with the naked eye; there is one system, the Andromeda Galaxy, which is more than two million light-years away, and is quite clearly visible as a tiny smudge against a dark background. We are seeing the Andromeda Galaxy not as it is today, but as it used to be more than two million years ago. If there are any astronomers in that system (as is highly probable), they will be seeing our Galaxy as it used to be before the start of the last Ice Age.

How large is the universe? Frankly, this is a question which is almost impossible to answer with confidence. The most remote object so far measured appears to be over 12,000 million light-years away. Whether the universe is infinite, or whether it has a definite size, is something which we can postpone for the moment.

I hope that this chapter has not been indigestible, but there was no choice but to cram it with facts. So to clarify matters, I will try to sum up the situation as concisely as possible:

1. The Earth is a planet, moving round the Sun at a mean distance of 93 million miles. The time taken to make one full circuit is, of course, one year – more accurately, 365¼ days.
2. The Sun is an ordinary star, appearing so glorious only because on the cosmical scale it is so close to us.
3. There are eight other planets moving round the Sun, of which five are easily visible with the naked eye. They have no light of their own, and depend upon reflecting the Sun's rays.

4. The Moon, at 239,000 miles, is much our nearest neighbour. It moves around us, and is much smaller than the Earth. Like the planets, it shines by reflected sunlight, which is why it shows its monthly phases. If the Sun were suddenly blotted out, the Moon and planets would disappear too, though the stars would be unaffected. (Let me assure you that nothing of the kind is likely to happen.)
5. The Sun, planets, planetary satellites and various bodies of lesser importance make up the Solar System.
6. The stars are suns in their own right, and are very remote. Even the nearest of them beyond the Sun is over four light-years away – a light-year being the distance travelled by a ray of light in one year, almost six million million miles.
7. Because they are so far away, the stars do not seem to move noticeably with respect to each other, so that the star-patterns or constellations do not change appreciably even over periods of centuries. The members of the Solar System, however, move around from one constellation to another, keeping to one definite band around the sky (the Zodiac).
8. The system of which our Sun is a member is known as the Galaxy. It contains about 100,000 million stars.
9. Beyond our Galaxy there are many other galaxies, most of which are many millions of light-years away.

I hope that this is helpful. And having cleared the basic facts, we can now turn our attention to the way in which the sky appears to move.

2

The revolving sky

We live upon a spinning globe. The Earth is rotating all the time, carrying us with it; a full turn takes 23 hours 56 minutes, conventionally rounded off to 24 hours. (Throughout this book I propose to deal mainly with rounded-off numbers, which makes things easier without being misleading.) We are not conscious of being whirled around, but the effects are obvious enough when we survey the sky. Since the Earth's direction of spin is from west to east, the sky appears to rotate from east to west, carrying the Sun, Moon, planets and stars along with it. (Most of the other planets spin in the same direction, though Venus rotates from east to west for reasons which remain a total mystery.)

Ancient stargazers believed that the stars were fixed on to an invisible crystal sphere, whose centre was concentric with the centre of the Earth. They called this the 'celestial sphere', and the term is still useful provided that it is not taken too literally. After all, the Earth does give the impression of being in the centre of the universe, with all the other celestial bodies revolving round it once a day.

The Earth's axis is tilted to the perpendicular to its path or orbit by 23½ degrees. It is this tilt which is responsible for the seasons; during northern summer the north pole is inclined towards the Sun, while in northern winter it is the turn of the south pole to be favoured. Actually, the Earth's orbit round the Sun is not quite circular. The distance from the Sun ranges from 91½ million miles in December to 94½ million miles in June, so that we are closest to the Sun when it is winter in London and New York. Still, the difference is not very much, and the effects are more or less masked by the fact

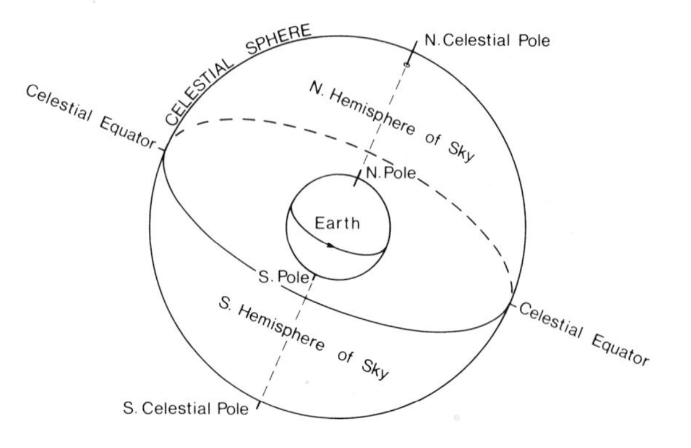

The celestial sphere

that there is so much ocean in the southern hemisphere; water tends to stabilize the temperature.

Now let us take a closer look at the so-called celestial sphere. In the diagram, the Earth is shown in the centre, with its two poles, its equator and its axis of rotation. If we prolong the axis line, we can fix the positions of the celestial poles. The north pole of the sky is marked within one degree by a fairly bright star Polaris in the constellation of the Little Bear. Therefore, Polaris seems to remain most stationary, with everything else moving round it in a period of twenty-four hours. This has nothing directly to do with Polaris itself, and neither has it always been the north pole star. The celestial pole shifts slightly over the years, and when the Egyptian Pyramids were being built the pole star was Thuban, in the Dragon. For the moment, however, and for centuries to come, Polaris occupies the position of honour. Navigators have always found it very useful, because its altitude above the horizon is equal to the observer's latitude on the surface of the Earth. Thus from London, where the latitude is 52 degrees, Polaris will be at an altitude of 52 degrees; from New York, latitude 40 degrees, Polaris will be only 40 degrees above the horizon (in round figures, that is to say). There is no bright star near the south pole of the sky, to the constant regret of southern navigators. The nearest naked-eye candidate is the obscure Sigma Octantis, which is so dim that even light mist will hide it.

Quite obviously, the south celestial pole will never be visible from a northern country, such as Britain. The solid body of the Earth gets in the way, so that the southernmost stars never rise above the horizon. On the other hand, the stars close to Polaris will never set, and will always be visible whenever the sky is sufficiently dark and clear. Ursa Major, the most famous of all constellations, comes into this category. (Note that in Britain Ursa Major is called the Great Bear or Plough; in the United States it is known as the Big Dipper.) It sweeps around the pole, and even when at its lowest it is still well above the horizon, so that it is said to be *circumpolar*.

The brilliant orange star Arcturus, which lies further south in the sky, is not circumpolar from Britain or the United States; at its lowest it drops below the horizon, so that it rises and sets regularly.

Anyone who travels from Britain towards the Earth's equator will notice that the altitude of Polaris is becoming less. Go to Mexico or central India, for example, and the angle between Polaris and the horizon will be reduced to

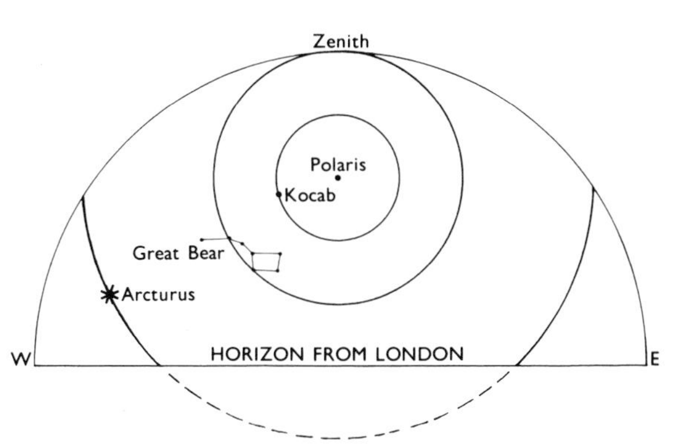

Kocab, a star not far from Polaris, describes a small circle. Ursa Major, further from Polaris, never sets, but can reach the zenith or overhead point. Arcturus can pass below the horizon, and so is not circumpolar.

only about 20 degrees. This means that Ursa Major will no longer be circumpolar, since part of its daily circle will lie below the horizon. To compensate for this, southern stars which never rise from Britain come into view – notably the brilliant Canopus, which is never to be seen from anywhere in Europe.

Go to the equator, and you will find that Polaris is right on the horizon; the south celestial pole lies on the opposite horizon, and the celestial equator passes overhead, so that the entire sky is visible at one time or another. In the Earth's southern hemisphere Polaris is lost to view, and from part of Australia and all of New Zealand even Ursa Major has gone. However, in its place are brilliant constellations such as the Southern Cross, which is actually much brighter than Ursa Major although it is smaller, and is shaped more like a kite than a cross.

The members of the Solar System take part in the daily rotation of the sky, but have individual movements of their own. The Sun seems to travel right round the sky in one year. Its light completely drowns that of the stars, although accurately pointed telescopes can show bright stars at any time. (The only time that stars can be seen with the naked eye during the daytime is when the Moon covers up the Sun, at a total solar eclipse.) The Sun's yearly path among the stars is known as the ecliptic. The belt of sky centred on the ecliptic is called the Zodiac, and it is here that the Sun, Moon and bright planets are always to be found.

The reason for this is that the orbits of the planets round the Sun are in much the same plane, so that if you draw a plan of the Solar System on a flat piece of paper you are not very far wrong. The inclination amounts to 7 degrees for Mercury and less than 4 degrees for all the other planets (apart from Pluto, which is much too faint to be seen with the naked eye or even a small telescope). The diagram again shows the celestial sphere, this time with the ecliptic. You can see that planets may appear in directions A or B, but never towards C or D. For example, you will never find a planet anywhere near Ursa Major or the Southern Cross.

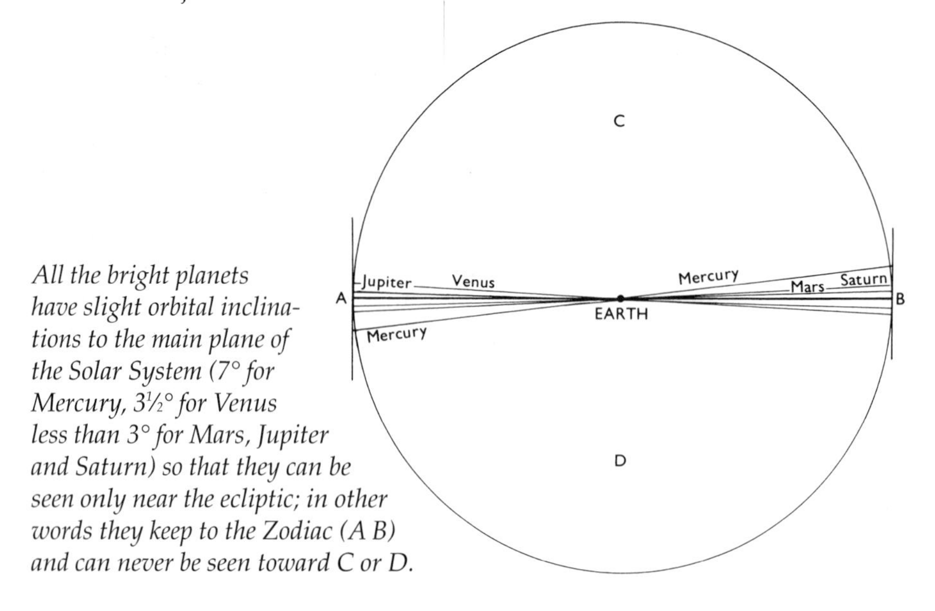

All the bright planets have slight orbital inclinations to the main plane of the Solar System (7° for Mercury, 3½° for Venus less than 3° for Mars, Jupiter and Saturn) so that they can be seen only near the ecliptic; in other words they keep to the Zodiac (A B) and can never be seen toward C or D.

Since it is impossible to put the planets into permanent star-maps, the observer has to learn how to recognize them – which is easy enough. Venus and Jupiter are always so brilliant that they far outshine any of the stars. Mars is noted for its strong red colour. Saturn, admittedly, can be confused with a star, but it is a slow mover, and once it has been identified it can be found again with no trouble. Mercury need not concern us for the moment, because it is never conspicuous, and the casual observer is unlikely to notice it at all. (Have you ever seen Mercury? In more than nine cases out of ten, I would be prepared to wager that the answer is 'No'.)

It has often been said that stars twinkle, while planets do not. This is not entirely true, but there is no doubt that a star, which is to all intents and purposes a point source of light, twinkles more than a planet, which shows as a small disk. Twinkling has nothing directly to do with the stars; it is due entirely to the Earth's unsteady atmosphere, which, so to speak, 'shakes' the starlight around. When a star is close to the horizon, its light comes to us through a deep layer of atmosphere, and the twinkling is pronounced; when the star is high up there is less atmosphere to be crossed, and the twinkling is much less marked. You can check this on any clear night by comparing the twinkling of stars at different heights above the horizon.

Apart from the circumpolar groups, the stars are seasonal. Thus Orion, one of the most brilliant of all the constellations, cannot be seen around June, because it is then too near the Sun in the sky and is above the horizon only during the hours of daylight. The monthly maps given in this book will, I hope, show exactly when and where the constellations are on view. Obviously I have had to give two main sets of maps – one for the northern hemisphere and one for the southern.

Most of this chapter has been concerned with the revolving sky, and there is an easy way to demonstrate it. Take an ordinary camera, wait until the sky is clear and dark, load with a fast film and leave the shutter open for a few minutes (or, if you like, an hour or two – though beware of 'fogging' the film by moisture or artificial lights). The stars will appear as trails as they cross the field of view. If you aim the camera at the north celestial pole, you will see that the trails are circular – and that Polaris itself shows a very short, curved trail, proving that it is not exactly at the polar point.

Finally, let me dispose of another common misunderstanding. It has been claimed that you can see stars in the daytime if you go down to the bottom of a mine-shaft, or a deep well, and look up at the narrow circle of light above you. In fact this is quite wrong. Stars are invisible in the daytime because the contrast between their light and the sky background is too low, and the situation to an observer in a coal-mine is exactly the same. No; if you want to see the stars during daylight, you must wait for the fleeting moments of a total eclipse of the Sun.

3

Patterns of stars

The first thing you will note when looking up at the night sky is that the stars seem to form definite patterns. Early men were suitably impressed, and divided the stars into groups, or constellations. Various systems were used; the Egyptians drew up one set of constellations and the Chinese another, but the system we use today comes from the Greeks. (Nobody is quite sure where it originated. There has been a suggestion that the Greek system came from the island of Crete, where the so-called Minoan civilization, named after the legendary King Minos, was highly developed until it was destroyed by a violent volcanic eruption around 1500 BC. Incidentally, there is little doubt that it was this disaster which gave rise to the story of Atlantis.)

Ptolemy, the last great astronomer of ancient times, listed forty-eight constellations, all of which are still to be found on our modern maps even though their boundaries have been modified. The names of the constellations were drawn partly from mythology and partly from everyday life. Thus we find Orion, the great Hunter; Hercules; and Pegasus, the flying horse which carried the hero Bellerophon into battle against a conventional fire-breathing monster, the Chimaera. We also find a Triangle, a Wolf and an Altar. However, Ptolemy lived in Alexandria, and could not see the stars of the far south, which were tackled later; some of the names have a decidedly modern flavour – the Telescope, the Microscope and so on. Moreover, there was a period when astronomers took a fiendish delight in tinkering with the already cumbersome and complicated constellations, adding such atrocities as Sceptrum Brandenburgicum (the Sceptre of Brandenburg), Officina Typographica (the Printing Press) and Globus Aerostaticus (the Balloon). Finally, in 1932, the controlling body of world astronomy, the International Astronomical Union, lost patience and reduced the accepted number of constellations to eighty-eight.

The constellations are of very different shapes, sizes and importance. The largest of all is Hydra, the Watersnake, with an area of 1303 square degrees; the smallest is the Southern Cross, with only 68 square degrees. Some constellations contain many bright stars, while others are so dim and formless that they seem to be unworthy of separate identity. However, the system has become so well established that it will certainly not be altered now. Periodic attempts have been made, perhaps the worst of which was to rename the constellations after prominent politicians. Mercifully, this idea was received with a total lack of enthusiasm.

In astronomy, the Latin names of the constellations are used; Latin is still the universal language, even though it has not been used in ordinary conversation for many centuries. The Latin names are easy enough, and I propose to use them here. A full list, with the English equivalents, is given on p. 196.

The individual star names are mainly Arabic, so that we have a truly international flavour: Greek constellations, Latin constellation names and Arabic star names (though some of the star names are Greek; Sirius is one). By now, only the names of the very brightest stars are used, and it would indeed be a daunting task to memorize individual names for even the few thousand stars which are visible with the naked eye.* In 1603 the German astronomer Johann Bayer proposed a different system, which was much more convenient and which is still used.

What Bayer did was to take each constellation separately, and give its stars Greek letters. The first three letters of the Greek alphabet are Alpha, Beta and Gamma; thus the brightest star in the constellation would be Alpha, the second brightest Beta, and so on down to the last letter of the alphabet, Omega. The Greek letter is followed by the genitive of the constellation name concerned, so that the brightest star in Cygnus, the Swan, becomes Alpha Cygni; the leader of Canis Major, the Great Dog, is Alpha Canis Majoris, and so on. Obviously the system is limited because there are only twenty-four letters, and a later astronomer, John Flamsteed (the first Astronomer Royal), gave the stars catalogue numbers. Thus Deneb is also known as 50 Cygni.

The Greek alphabet is easy to learn. It is as follows:

α Alpha	ε Epsilon	ι Iota	ν Nu	ρ Rho	φ Phi
β Beta	ζ Zeta	κ Kappa	ξ Xi	σ Sigma	χ Chi
γ Gamma	η Eta	λ Lambda	ο Omicron	τ Tau	ψ Psi
δ Delta	θ Theta	μ Mu	π Pi	υ Upsilon	ω Omega

The next important thing to note is the star's apparent brightness, which is a measure of its magnitude. The scale works in the manner of a golfer's handicap, with the more brilliant performers having the lower magnitudes. Thus very bright stars are of magnitude 1; magnitude 2 is fainter, 3 fainter still, and so on. Stars below magnitude 6 are normally invisible with the naked eye even on a dark night. Starting from zero, the magnitudes are roughly as follows:

0: extremely bright stars, such as Capella in Auriga (the Charioteer) and Vega in Lyra (the Lyre).
1: Very bright stars standing out against their neighbours. Conventionally, any star brighter than magnitude 1.5 is said to be of the 'first magnitude'; there are only twenty-one of them.
2: Moderately bright stars, such as Polaris.
3: Fainter stars, still easy to see even in conditions of mist or fairly strong moonlight.
4: Fainter still, concealed by moonlight.

* Beware of unofficial 'agencies' which claim to be able to name stars – naturally for a fee! These agencies are entirely bogus. Have nothing to do with them.

5: Too faint to be seen with the naked eye except when the sky is dark and clear.
6: Faintest stars visible with the naked eye under good conditions.

The scale is extended both ways. Venus, the most brilliant planet, has a magnitude of about minus 4, while on the other hand the world's largest telescopes can record objects as dim as magnitude plus 30. Magnitudes can be measured very accurately with special devices known as photometers, but the naked-eye observer will be well satisfied with a value given to the nearest tenth. For instance Alpha Lupi, the leader of Lupus (the Wolf) is of magnitude 2.3, rather fainter than Polaris at 2.0. Four stars have magnitudes below zero: Sirius (−1.5), Canopus (−0.7), Alpha Centauri (−0.3) and Arcturus (−0.04). On the same scale, the Sun's magnitude is almost −27.

Incidentally, there is more sense to these magnitudes than might be thought. The measurements have been devised according to a logarithmic scale. Thus a star of magnitude 1.0 is exactly a hundred times as bright as a star of magnitude 6.0.

Note that magnitude is a measure of a star's apparent brightness, not its real luminosity. The stars are at very different distances from us, and a good example can be given by comparing Sirius, in Canis Major, with Rigel in Orion. As we have noted, the magnitude of Sirius is −1.5; that of Rigel is 0.1, so that Sirius is well over a magnitude the brighter of the two. Yet Sirius is a mere 8.6 light-years away, and has 26 times the power of the Sun; Rigel, at around 900 light-years, is the equal of 60,000 Suns put together. If Rigel were as close to us as Sirius, it would cast shadows. In astronomy, appearances can often be highly deceptive.

It follows that what we call a 'constellation' is nothing more than a line-of-sight effect. Consider Cygnus, the Swan, whose five stars make up an X-form which has led to the constellation being nicknamed the Northern Cross. The five stars are:

	Magnitude	Distance in light-years
Alpha (Deneb)	1.2	1800
Gamma	2.2	750
Delta	2.9	160
Epsilon	2.5	81
Beta (Albireo)	3.1	390

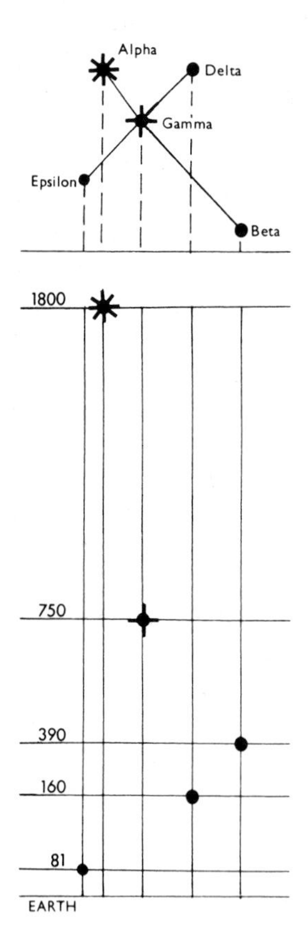

Distances from Earth in light-years of the five stars in the Cross of Cygnus

Note, incidentally, that the Greek letters are out of order: Beta Cygni should be the second brightest star in the constellation, not the fifth. This often happens. The proper name of Beta, Albireo, is often used because it happens to be a particularly beautiful double star, with a yellow primary and a blue companion – though with the naked eye it appears single.

Now let us make a chart, showing the five stars at their correct relative distances. From Earth, they make up the X-pattern; but if we happened to lie in a position in space between, say, Deneb and Gamma, then Deneb would be on one side of the sky and the other four stars on the opposite side, so that the familiar X would be lost. Another good example is provided by Alpha and Beta Centauri, the brilliant pointers to the Southern Cross (never visible from Europe). Alpha Centauri – which, surprisingly, has no accepted proper name – is the nearest of all the bright stars, at 4.3 light-years from us; Beta Centauri (Agena) is 460 light-years away. The two have absolutely no connection with each other, even though they happen to lie side by side in the sky as seen from Earth.

This, of course, is yet another proof of the absurdity of astrology – I promise that I will mention it no more! Astrologers believe that the constellation names have real significance; thus Pisces, the Fishes, is a 'watery' sign. In fact Pisces is made up of a few faint stars which do not give the slightest resemblance to the outline of fishes or anything else, and if we had followed the Chinese or the Egyptian system there would be no Pisces at all. Neither would there be a Great Bear, though on the Egyptian pattern we would have a Cat and a Hippopotamus.

The planets and the Sun and Moon, are limited to the Zodiac, which stretches to either side of the ecliptic. There are twelve Zodiacal constellations, ranging from Aries (the Ram) through to Pisces, though a thirteenth constellation, Ophiuchus (the Serpent-Bearer) does cross the Zodiac at one point. Again there are great differences in size and importance between the constellations in the Zodiac: Scorpio or Scorpius (the Scorpion) and Gemini (the Twins) are large and brilliant, while Cancer (the Crab) and Libra (the Balance) are as dim and formless as Pisces. The constellation patterns are man-made, and are entirely arbitrary.

To my mind, the best way of learning the sky is to take a few constellations which cannot possibly be missed, and use them as direction-finders. Ursa Major, the Great Bear, is particularly useful to observers in the northern hemisphere. Its seven 'Plough' stars are circumpolar in Britain, and can be pressed into service. The two end stars of the pattern (Merak and Dubhe) point to Polaris, the Pole Star, in Ursa Minor (the Little Bear); the three end stars of Ursa Major point to Arcturus, the brilliant orange star in Boötes (the Herdsman) and then to Spica in Virgo (the Virgin) – and so on, though of course Arcturus and Spica are too far south to be circumpolar from Britain or North America. Orion is even more useful when it is visible. The three bright stars of the Hunter's Belt point one way to Sirius and the other way to Aldebaran, a bright orange star in Taurus (the Bull).

The stars, then, are by no means alike; and before going on to describe the constellations in more detail, I feel that it will be helpful to say more about the nature of a star, and how it evolves throughout its long lifetime. First, however, I propose to make a diversion and recount some of the legends surrounding the stars.

4

Star legends

I have always been fascinated by mythology, particularly Greek mythology. There is something curiously attractive about the old Olympians and their doings, even though so many of the heroes came to unpleasant ends. And because some of the legends are commemorated in the sky, it seems worth breaking away temporarily from hard facts and figures to say more about them, bearing in mind that many of the legends have several different versions.

Take the two Bears, Ursa Major and Ursa Minor. According to mythology, Ursa Major was originally a beautiful maiden, Callisto, daughter of King Lycaon of Arcadia, who was attendant to the goddess Juno. Sadly for Callisto, she was decidedly more lovely than Juno, and this was something that the goddess did not like at all, so she ill-naturedly turned Callisto into a bear. Years later Callisto's son Arcas met the bear when he was out hunting and, not realizing that it was his loving mother, prepared to shoot it. At this juncture Jupiter,* king of the gods, stepped in. He hastily turned Arcas into a bear also, and then reached down, swinging them skyward by their tails and

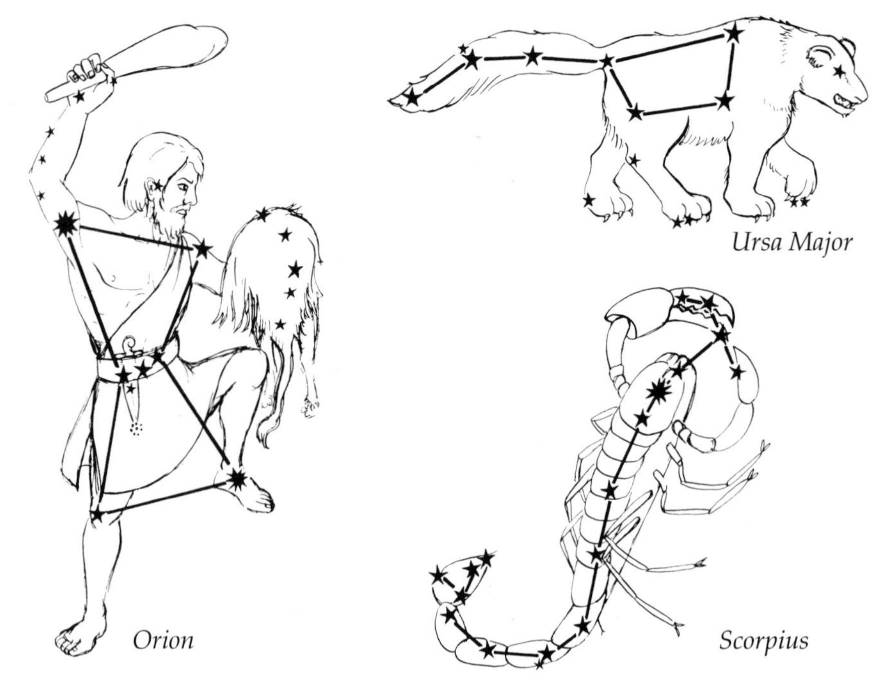

placing them among the stars. This is one possible reason why both the Bears have exceptionally long tails!

Then there is Orion, the mighty hunter, who is said to have been a son of the sea-god Neptune. Orion boasted that he could kill any creature on Earth. However, Juno, who had reason to be jealous of him (Juno often appears in a rather poor light in these legends), called in a scorpion, which crawled out of a hole in the ground, bit Orion in the heel and brought his career to an abrupt conclusion. Subsequently Diana, goddess of hunting, pleaded with Jupiter to bring Orion back to life and set him in the sky. This was duly done. Just to be fair, the scorpion also was elevated to celestial rank, but was tactfully placed in the opposite side of the sky.

As befits a hunter, Orion has his retinue; Canis Major (the Great Dog) and Canis Minor (the Little Dog) are his companions, while Lepus, the Hare, was placed beneath his feet. Orion's shoulder is marked by the bright red star Betelgeux, and his foot by the even brighter Rigel. Above him in the sky (as seen from the Earth's northern hemisphere) is Auriga, the Charioteer, led by the brilliant yellow Capella. In legend, Auriga represents Erechthonius, a son of the blacksmith god Vulcan. Erechthonius was born deformed, and was brought up by the goddess Minerva without the knowledge of the other Olympians. When he reached manhood he became King of Athens, and invented the four-horse chariot, for which he was rewarded with a place in the sky.

Hercules, the mighty hero, is also to be found – though frankly he is not as brilliant as perhaps he ought to be. There are several of his victims on view: Leo (the Nemaean Lion which he had to kill during his famous labours), Hydra (the hundred-headed monster which lived in the Lernaean marshes until Hercules disposed of it) and Cancer (the Crab). Cancer was a gigantic sea-crab, which was sent by Juno (again!) to attack Hercules during his battle with the Hydra. Not unnaturally, Hercules trod on the crab and squashed it, but Juno, not to be thwarted, placed it in the sky, presumably in recognition of 'services rendered'.

Ophiuchus, the Serpent-Bearer (called Serpentarius on some of the old maps) represents Aesculapius, a son of Apollo, who became so skilled in medicine that he not only healed the sick, but even restored the dead to life. This alarmed Pluto, god of the Underworld, who protested that his kingdom would be in grave danger of being depopulated if this sort of thing went on. Rather reluctantly Jupiter struck Aesculapius down with a thunderbolt, but, of course, a place in the sky for him was automatic.

There is a charming story attached to the little northern constellation of Delphinus, the Dolphin. It concerns Arion, a singer whose voice was so beautiful that he won any competition in which he entered. On one occasion he was sailing home to Sicily from Corinth, laden with prizes, when the ship's crew decided to throw him overboard and appropriate the prizes for themselves. However, Arion was saved by a dolphin, which carried him to the shore and took him safely to the port of Jaenarius. When the dolphin died, at an advanced age, its benevolent act was properly rewarded.

* Jupiter is the Roman name for Zeus, the Greek ruler of the gods, but it seems best to use the more familiar names here.

Aries, the Ram, is usually regarded as the first constellation of the Zodiac (though for reasons which need not concern us at the moment, it should now be ranked second). It is said that Athamas, King of Thebes, had two children, Phryxus and Helle. The children's stepmother, Ino, had every intention of killing them, and so Mercury, the messenger of the gods, sent a flying ram to rescue them. For good measure, the ram's fleece was made of gold. Unfortunately, Helle fell off the ram as they were soaring over the ocean (the site of her watery grave is still known as the Hellespont), but Phryxus reached safety. After the ram died, its fleece was placed in a sacred grove, guarded by a dragon – where it stayed until it was forcibly removed by Jason and his comrades, the Argonauts.

Jason has no place in the sky (I have often wondered why not), but maps used to contain his ship, the great *Argo*. However, the constellation of Argo Navis was so huge and unwieldy that in the 1932 revision, the International Astronomical Union committee chopped it up into a keel (Carina), a poop (Puppis) and sails (Vela). Thus Canopus, which used to be known officially as Alpha Argûs, has become Alpha Carinae.

Virgo, the Virgin, was identical with another of Jupiter's daughters, Astraea, the goddess of justice, who ruled the world during its Golden Age. When the standards of behaviour deteriorated, Astraea became so disgusted that she decided to go home to Olympus.

Probably the most famous of all the legends is that of Perseus, killer of the Gorgon, Medusa, whose hair was made of snakes and whose glance could turn anyone to stone. Perseus had what may be regarded as an unfair advantage, since the gods had provided him with winged sandals and also a shield which he could use to locate Medusa, looking only at her reflection and not at the Gorgon herself. Having decapitated Medusa, Perseus was returning home when he saw the beautiful Princess Andromeda tied to a post on the seashore. This was clearly something worthy of investigation. In fact Andromeda's mother, Queen Cassiopeia, had fallen foul of the sea-god Neptune, and a monster had been sent to ravage the country. King Cepheus consulted the Oracle, and was told that the only way to placate Neptune was to sacrifice his daughter to the monster. Fortunately Perseus came along in the nick of time and turned the beast to stone. He then married Andromeda, thereby providing one of the rare cases of a legend with a happy ending. Perseus, Cepheus, Cassiopeia, Andromeda and even the sea-monster (Cetus) are all on view, though Cetus is often relegated to the status of a harmless whale.

The modern constellations do not have mythological associations; one can hardly find much romance in, say, an air-pump or a microscope. There are a few constellations which bear at least a passing resemblance to the objects which they represent; such are the Triangle and the two Crowns. It is also possible to visualize a scorpion in the long line of bright stars making up Scorpius. In most other cases one needs a vivid imagination to find any resemblance – but no matter; no doubt it would be easy enough to work out an orderly and clinical system, but it would be a pity to lose our mythological legends.

Now let us turn back to true science, and see what we can find out about the stars themselves.

5

The stars themselves

The Sun is the only star sufficiently close to us to be studied in real detail – the rest appear as mere points of light – and therefore 'solar physics' is the vital key to our understanding of all the other stars.

First, it is wrong to assume, as many people do, that the Sun is on fire. A Sun made up entirely of coal, and blazing furiously enough to give out as much heat as the Sun actually does, would not last for long on the cosmical scale; but we know that our Earth is at least 4600 million years old, and the Sun must date back further than that. We now know that the Sun is shining by nuclear reactions taking place deep inside it, where the temperature is of the order of 14 million degrees Centigrade (perhaps more) and the pressures are colossal. Hydrogen is the most plentiful substance in the universe, and the Sun contains a great deal of it. In the solar core, the nuclei of hydrogen atoms are banding together to form nuclei of helium atoms. It takes four hydrogen nuclei to make one helium nucleus; each time this happens, a little energy is set free and a little mass is lost. It is this energy which keeps the Sun shining, and although the mass-loss amounts to 4 million tons per second there is no reason to believe that the Sun will change much for several thousands of millions of years in the future.

Eventually the supply of available hydrogen will be used up, and the Sun will change its structure; the inside will shrink and heat up still further, while the outer layers will expand and cool. The Sun will become a Red Giant star, as Arcturus in Boötes is now – and this will mean the end of all life on Earth, if not the end of the Earth itself, because there will be a period when the Sun will radiate at least a hundred times as powerfully as it does now. This will not last for very long. The Sun will shed its outer layers, and what is left will be a very small, very dense star of the kind known as a White Dwarf. Finally all the Sun's light and heat will disappear, and the end result will be a cold, dead globe still circled by the ghosts of its remaining planets.

Stars which have much the same mass as the Sun will follow similar life-stories. A star which is much less massive will never become hot enough inside for nuclear reactions to begin, so that it will shine feebly for a while and then die out. A star which is much more massive than the Sun may explode in a blaze of glory known as a supernova outburst; in extreme cases the star will become so dense and so small that not even light will be able to escape from it. The old, collapsed star will be surrounded by a sort of forbidden zone, cut off from the outside universe: a Black Hole, in fact.

Meanwhile, what can we learn about the stars from sheer naked-eye observation?

First, the stars are not all of the same colour. The Sun, as anyone can see, is yellow; a star with a cooler surface will be orange or orange-red, while a hotter star will be white or bluish. The very hottest stars have surface temperatures of well over 50,000 degrees Centigrade, while at the other end of the scale the red stars are no hotter than 2000 or 3000 degrees. The Sun has a surface temperature of almost 6000 degrees.

Use a telescope, or even binoculars, and these diverse colours will be well seen; for instance, Vega, almost overhead from northern latitudes in midsummer, is strongly blue, while Betelgeux is orange-red. Unfortunately, not many of these colours are obvious with the naked eye. All that can really be said is that a few stars are orange or red, while some of the brighter ones have a bluish cast; the rest simply look white.

Of the special kinds of stars, the most spectacular are the doubles. The best example is Mizar, in Ursa Major. When the sky is dark, Mizar is seen to have a much fainter star, Alcor, close beside it. With a telescope, Mizar itself may be split into two components, so close together that with the naked eye they appear as one.

Many thousands of double stars are known. Some of them are due to line-of-sight effects, since if one star lies almost behind the other, as seen from Earth, the two will appear side by side in the sky. Yet oddly enough, these non-related or 'optical' pairs are in the minority. Usually, the two components of a double star are genuinely associated, and are moving round each other much as the two bells of a dumb-bell will do when twisted by their joining bar. A system of this kind is known as a 'binary'. If the components

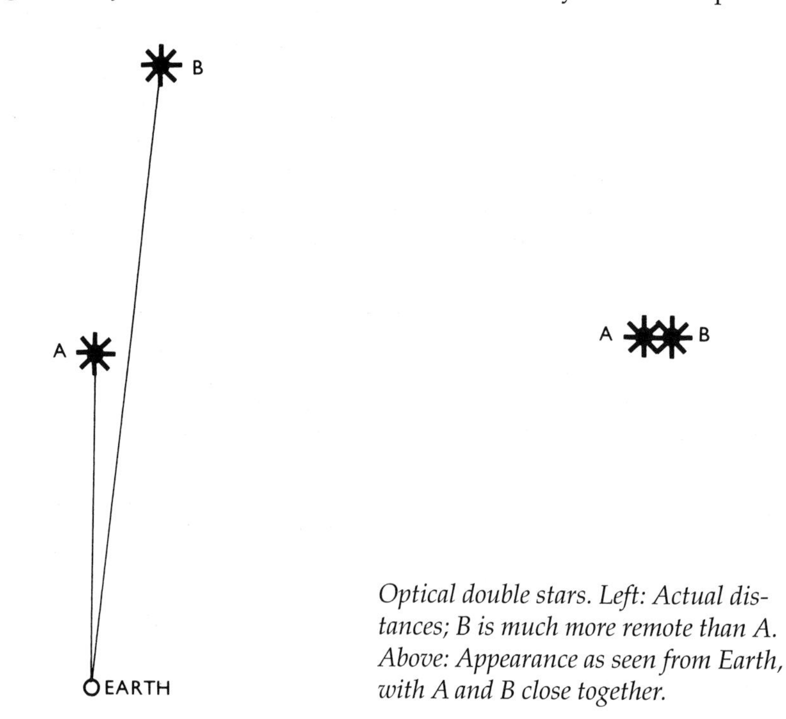

Optical double stars. Left: Actual distances; B is much more remote than A. Above: Appearance as seen from Earth, with A and B close together.

are close together, it will take only a few hours for the two stars to complete a full revolution; if the components are wide apart, the revolution period may amount to millions of years.

We are not concerned here with telescopic binaries, and, obviously, all double stars which are wide enough to be split with the naked eye must be either optical pairs, or else binaries with very long periods. One odd fact which has emerged is that binary systems are remarkably common, and may even be the rule rather than the exception, so that they may outnumber single stars such as the Sun. Triple stars, quadruple stars and still more complex groups are known, and have been termed 'family parties' of stars; thus Castor in Gemini, the Twins, is made up of three pairs – two bright and one faint – even though with the naked eye it looks single. It used to be thought that a binary was produced by the break-up of a massive star, but this is now known to be wrong. The components of a binary (or a multiple star) are formed separately, at the same time and from the same gas-and-dust cloud in space.

Next we come to variable stars, which are of special interest to the naked-eye observer because one can watch them perform. They are very numerous, though it is true that not many of them become really bright.

Most stars shine steadily for year after year, century after century, but a variable – as the name implies – shows short-term changes; the 'period' denotes the interval between one maximum and the next. Variable stars are of different types. The so-called Mira stars, named after the best-known member of the class (Mira or Omicron Ceti, in the constellation of the Whale), have periods of from a few weeks to well over a year; the period of Mira itself is on average 331 days, though it is not quite constant and may range a week or two to either side of the mean value. The magnitude varies between 2 and 9, so that minimum Mira is too faint to be seen with the naked eye or even binoculars, while at its best it may surpass the Pole Star. Not all maxima are equal; sometimes the magnitude never exceeds 4, and generally Mira is visible with the naked eye for only a few weeks in every year.

The Mira stars are old Red Giants. There are also the semi-regular variables, such as Betelgeux, which have smaller ranges of magnitude and only rough periods, while other stars are completely irregular, so that one can never tell what they will do next.

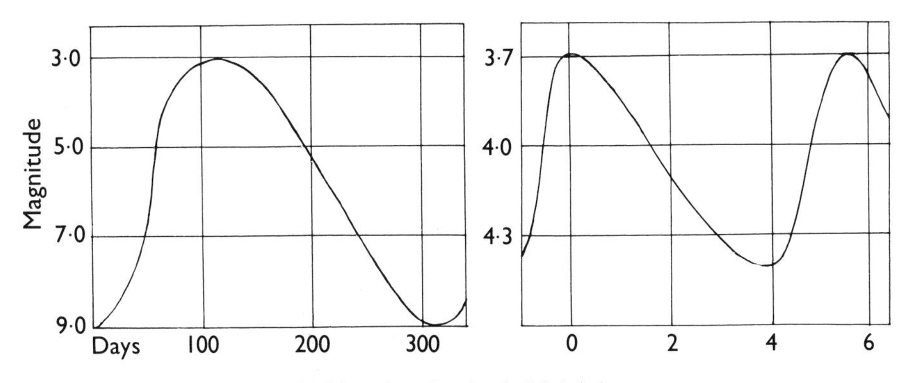

Light-curves of Mira Ceti (left) and Delta Cephei (right).

The Cepheid variables, named after the first discovered member of the class (Delta Cephei, in the far north of the sky) are as regular as clockwork. The periods are generally around a few days; 5.3 days for Delta Cephei itself. These Cepheids are very useful astronomically, because their periods are linked with their real luminosities, with the longer-period stars being the more powerful. Once a star's real brightness is known, its distance can be worked out, and so the Cepheids act as invaluable 'standard candles' in space. Luckily they are very luminous, and can be seen over tremendous distances.

Then there are the eclipsing variables – which should more properly be called eclipsing binaries, because they are not truly variable at all even though they show changes in brightness. Algol, in Perseus, is the prototype star. For most of the time it is of the second magnitude, but every two and a half days it begins to fade, dropping to below magnitude 3 in a few hours and remaining faint for twenty minutes before starting to recover. The key to this peculiar behaviour was found in 1783 by John Goodricke, who was deaf and dumb and who died at the early age of twenty-one. Goodricke realized that there are two Algols, one much brighter than the other. They move round their common centre of gravity over a period of two and a half days; when the fainter component passes in front of the brighter, part of the total light we receive is cut off. There is a secondary minimum when the brighter component eclipses the fainter, but with Algol this minor dip is too slight to be noticed with the naked eye.

Beta Lyrae, near Vega in the sky, is also an eclipsing binary, but here the two components are less unequal, so that apparent variations are always going on. No telescope will show Beta Lyrae as anything more than a speck of light, but we know that it is made up of a pair of egg-shaped stars almost in contact, and that the whole system is surrounded by swirling clouds and streamers of gas. Seen from close range, Beta Lyrae would be fascinating indeed.

Naked-eye observers can follow a number of variable stars, and estimate their magnitudes from night to night (even hour to hour with Algol and other eclipsing binaries). The procedure is to compare the variable with two neighbouring stars which are constant. For instance, take Delta Cephei, when the range is from magnitude 3.5 to 4.4. Suitable comparison stars are Zeta Cephei (3.3), Iota (3.7) and Epsilon (4.2). Suppose that we find Delta to be 0.2 magnitude fainter than Iota, and 0.3 brighter than Epsilon? Obviously, Delta's magnitude at this moment will be 3.9.

Trained eyes can distinguish differences of as little as a tenth of a magnitude, but there are various traps. In particular, a star which is low down will suffer 'extinction' – that is to say it will appear fainter than it really is, because its light is coming to us through a deep layer of the Earth's atmosphere. Therefore, it is wise to select comparison stars which are at about the same altitude as the variable. Also, many variables (the Mira stars and the semi-regulars) are reddish, and it is not easy to compare a red star with a white one. However, suitable precautions can be taken, and the observer can draw up a light-curve, with magnitude plotted against them.

Novae, or 'new stars', appear without warning. The name is misleading, because a nova is not in fact new. What happens is that a formerly dim star flares up to many times its normal brightness, remaining near maximum for a few days or a few weeks before sinking back to obscurity. I well remember going outdoors on the evening of 29 August 1975 and finding a star of almost the second magnitude in Cygnus, near Deneb, which had not been there on the previous evening. (Let me add that I was only one of dozens of independent discoverers. Prior claim went to the Japanese astronomer Honda, who found the nova several hours before darkness fell over Britain.) By 30 August it had risen to magnitude 1.8, but by the end of the first week in September it had dropped below naked-eye visibility, and it has now become very faint indeed.

According to current ideas, a nova is a binary system, made up of a normal star together with an old White Dwarf. The White Dwarf pulls material away from its larger but less dense companion, building up a shell or disk of material. Finally the temperature rises so much that nuclear reactions are triggered off, and there is a violent though temporary outburst.

Novae visible with a telescope are not uncommon, but few of these interesting objects become bright enough to be seen without optical aid. All the same, the naked-eye observer has always the chance of making an important discovery – and it is true that amateurs usually know the sky much better than professional astronomers! If you think you have happened upon a nova, check very carefully before reporting it, as it is only too easy to be deceived by a weather balloon catching the sunlight, or even a slowly moving artificial satellite. (Some years ago I had a telephone call from a professional astronomer who believed that he had found a bright nova in Libra. It turned out to be Saturn.)

Supernova outbursts, marking the death of a massive star, are much less frequent. No supernova has been seen in our Galaxy since 1604, before telescopes appeared on the scene. Astronomers regret this, since they would very much like to study a nearby supernova with modern equipment, but at least a supernova was seen in 1987 in the Large Cloud of Magellan, at a mere 169,000 light-years, and was intensively studied.

It used to be thought that a nova (or supernova) could be explained by the head-on collision of two stars. The idea seemed logical enough – until it became possible to measure the size of the Galaxy. In fact the stars are so widely separated in space that collisions must be very rare indeed.

Yet we do see genuine groups of stars. These are termed clusters, a name which sums them up perfectly. The most famous of them is known popularly as the Seven Sisters, though officially as the Pleiades. It lies in Taurus (the Bull) and is prominent with the naked eye; at first sight it looks like a hazy patch, but normal-sighted people can make out at least seven separate stars, and binoculars will show many more. Other celebrated open or loose clusters are the Hyades, round Aldebaran; Praesepe or the Beehive, in Cancer; and the lovely Jewel Box in the Southern Cross. The stars in such a cluster have a common origin, and were born at around the same time in the same gas-and-dust cloud in space.

Very different are the globular clusters, whose outlines are much more regular. With the naked eye they look circular; telescopes will show separate stars near their edges, but near the centres the crowding is much greater, and a large instrument is needed to resolve the whole of the glowing mass into stars. Globular clusters lie near the edge of the Galaxy, and so are a long way off – over 15,000 light-years in most cases – so that they look faint. Only three are clearly visible with the naked eye. Of these, one is in Hercules. The others are in the far south, never visible from Europe: one in the constellation of the Centaur and the other in the Toucan.

If our Sun, with its family of planets, were a member of a globular cluster, the night skies would be glorious. There would be many stars only light-months away, instead of light-years, and they would be brilliant enough to cast shadows; true darkness would be unknown. Many of these stars would be orange or red, because globular clusters are very ancient, and their leaders have evolved into the Red Giant stage or beyond.

Nebulae come into another category altogether. The Latin word nebula means 'cloud', and the dim, misty patches of light in the night sky known as nebulae do look just like faint, luminous clouds. One splendid example lies in the Sword of Orion, which extends away from the three bright stars of the Hunter's Belt. It is easy to see with the naked eye, and has been known for centuries. It shines because it is being lit up by very hot stars mixed in with it.

Nebulae are stellar birthplaces; inside them, fresh stars are being born from the gas and dust. Several nebulae are visible with the naked eye, though unfortunately most are too faint to be seen without optical aid.

Until less than eighty years ago, all nebulae were thought to be members of our Galaxy. Then came a spectacular discovery. The gas-and-dust nebulae, such as in Orion's Sword, do belong to our Galaxy, but other nebular objects proved to be galaxies in their own right. Such is the Great Spiral in Andromeda, which is over two million light-years away, and contains more than our own quota of 100,000 million stars. It is on the fringe of naked-eye visibility; on a really dark and clear night it can just be located as a dim blur. Binoculars, of course, show it clearly, though photographs taken with large telescopes are needed to bring out its spiral structure. Moreover it lies at an unfavourable angle to us, so that the full beauty is rather spoiled.

External galaxies are bewilderingly numerous; at least 1000 million are within our range, and they contain objects of all kinds, from normal stars to giants, variables, novae, clusters and even occasional supernovae. Yet because they are so remote, only three are visible with the naked eye – at least to people with average sight. The Andromeda Spiral is one; the others are the two Clouds of Magellan, which are too far south to rise over Europe or the United States. The Clouds are often regarded as satellite systems of our Galaxy; the Large Cloud is so bright that it can be seen even in moonlight Their distances are less than 200,000 light-years.

The Andromeda Spiral, the Clouds of Magellan, and the spiral in Triangulum (which some keen-sighted people claim to see with the naked eye) belong to what is termed the Local Group of galaxies. Beyond this Local Group, all the galaxies are racing away from us, so that the whole universe is expanding. The most remote objects so far discovered seem to be over 12,000 million light-years away, so that we are now seeing them as they used to be long before the Earth came into existence.

In 1781 the French astronomer Charles Messier compiled a list of star-clusters and nebulae. Ironically, he was not interested in them; he was a comet-hunter, and listed the clusters and nebulae as 'objects to avoid', though

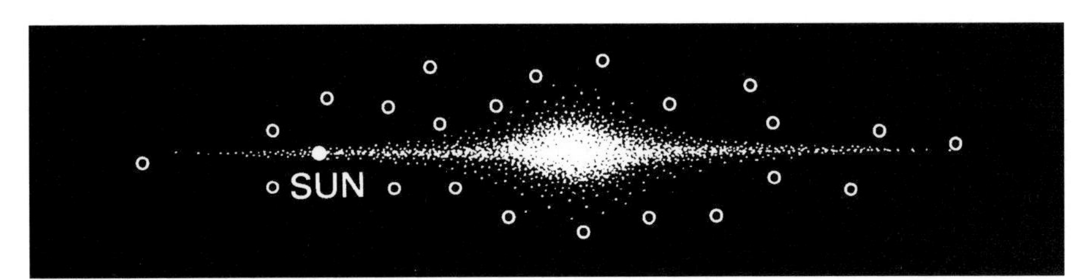

Position of the Sun in the Galaxy. Globular clusters are represented by open circles.

today we still use Messier's numbers, and his comet discoveries have been more or less forgotten! Thus the Andromeda Spiral is Messier 31, or M. 31; the Orion Nebula is M. 42, Praesepe M. 44, and so on. The recent Caldwell catalogue, now widely used, contains over 100 bright nebular objects not listed by Messier.

Finally, nobody can fail to notice the lovely band of the Milky Way. It is yet another line-of-sight effect, seen when we look along the main plane of the Galaxy. Here and there we can make out dark, starless patches, which are not genuine 'holes' but are simply dark nebulae, unlit by any stars mixed in with them and detectable only because they blot out the light of stars beyond. The best example is the Coal Sack in the Southern Cross, though there are others in Cygnus and elsewhere.

This description of the stellar sky is of necessity incomplete, but for the moment it will suffice. Now let us turn to the night sky as it appears at different seasons of the year, beginning with the view as seen from the northern hemisphere of the Earth.

6

Stars of the far north

It cannot be said that the stars near the north pole of the sky are either particularly brilliant or particularly interesting, but at least there are several groups which are extremely easy to identify. From Britain, and from the northern United States, both the Bears, Ursa Major and Ursa Minor, as well as Cassiopeia are circumpolar, so that they never drop below the horizon. The Bears skate along the horizon of most of South Africa and part of Australia, but from New Zealand they are out of view altogether. Of course, the Pole Star never rises south of the Earth's equator.

In drawing up charts to be used for star recognition, the only possible method is to select one definite latitude for the observer and design the maps accordingly. For the northern hemisphere I have chosen latitude 50° N, which applies to the southernmost part of England and the northernmost part of the United States. In point of fact, outline maps are not badly affected by minor changes in the observer's latitude; these will do quite well for the whole of Europe and a large part of the North American continent. For all practical purposes, they apply to anyone living south of Iceland and north of

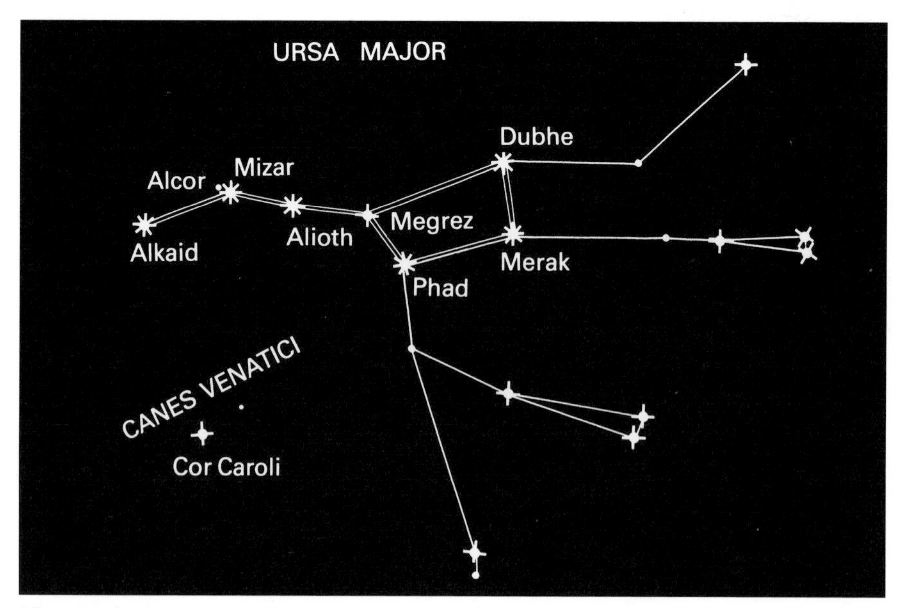

Ursa Major

Mexico. There are certain obvious differences – for instance, the 'sting' of the Scorpion is excellently seen from Arizona, but not from England – but these can be allowed for.

URSA MAJOR: *the Great Bear*

Ursa Major is a large constellation. The so-called Plough, made up of seven stars, is only part of it, and the name is unofficial; it has also been called King Charles's Wain, while Americans know it as the Big Dipper. On the whole, we must admit that the Americans are the most rational. The pattern is nothing like that of a bear or a plough, but it does give a slight impression of a spoon or ladle.

Ursa Major is shown here, together with the rest of the constellation. The proper names of all seven stars have been given, because they will be used over and over again as guides to other constellations. Starting at the end of the plough handle (or the spoon handle), the stars are:

Eta or Alkaid (magnitude 1.9)
Zeta or Mizar (2.1)
Epsilon or Alioth (1.8)
Delta or Megrez (3.3)
Gamma or Phad (2.4)
Beta or Merak (2.4)
Alpha or Dubhe (1.8)

The proper names are Arabic, and some of them have variants; thus Phad has also been called Phekda or Phecda, while Alkaid has the alternative name of Benetnasch.

Ursa Major is extremely easy to find, not because its stars are brilliant, but because the pattern is so well marked. Broadly speaking it may be seen rather low in the north during winter evenings, high in the north-east during spring evenings, high in the north-west during summer evenings, and descending in the north-west during autumn evenings. It may pass right overhead, as it does around midnight in March.

It may, of course, appear at all sorts of angles. Look for it during a late evening in January, and you will see the Bear 'standing on its tail', as shown in Star Map 1 (see p. 35). But wherever it may be, it stands out at once.

Alkaid, Dubhe and Alioth are all very slightly above the second magnitude, and to all intents and purposes they appear equally bright, but this does not mean that their true luminosities are the same. Alkaid is over 100 light-years away, Dubhe and Alioth less than 80, so that Alkaid is really the most powerful of the three.

Megrez, or Delta Ursae Majoris, is obviously fainter than its six companions, and there is a minor mystery about it, because the old astronomers of more than a thousand years ago stated that it was as bright as the other Ursa Major stars. Either they were wrong, or (less probably) Megrez has actually faded by about a magnitude since then.

With one exception, all Ursa Major stars are white. The exception is Dubhe, which is somewhat orange in hue, showing that its surface temperature is lower. The difference is not hard to detect with the naked eye; binoculars, of course, bring it out very clearly.

The most famous member of Ursa Major is Mizar, because it is accompanied by the fourth-magnitude Alcor. The Arab astronomers of a thousand years ago were familiar with it, but, strangely, they regarded it as a test of eyesight, which is certainly not true today; it is easy to detect against any reasonably clear and dark sky, though haze will conceal it.

Apart from Alkaid and Dubhe, the stars of Ursa Major are travelling through space in much the same direction at much the same rate, so they make up what is termed a 'moving cluster'. The individual movements are too slight to be noticed with the naked eye over periods of many lifetimes, but eventually the shape will become distorted; 50,000 years hence, for example, Alkaid will lie 'below' Mizar, and Dubhe will have made off away from Merak.

Undoubtedly Ursa Major is the most useful signpost in the sky to Britons and North Americans. Moreover, the position of Ursa Major itself will indicate how many of the surrounding groups are on view and how many are not. When Ursa Major is standing upright over the horizon, for instance, there is no point in looking for Arcturus, which will be out of sight.

The rest of Ursa Major covers a wide area, but is not particularly conspicuous. There are two triangles, made up of stars between magnitudes 3 and 4, easy to identify when the constellation is high in the sky but often obscured by horizon mist when Ursa Major is at its lowest.

URSA MINOR: *the Little Bear*

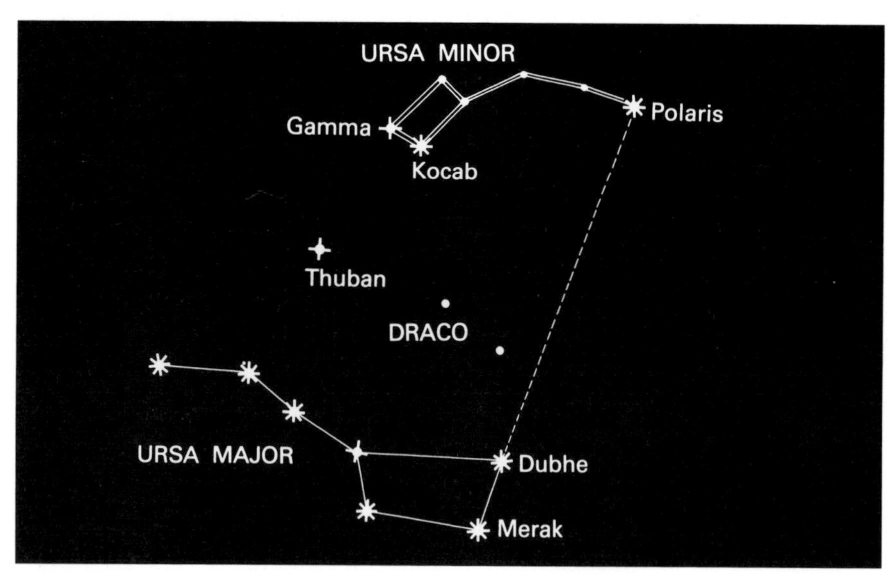

Ursa Minor

Once Ursa Major has been found, there should be no difficulty in locating Ursa Minor, the Little Bear – or, as Americans call it, the Little Dipper. By pure coincidence, the shape of Ursa Minor is not unlike that of Ursa Major, though the individual stars are fainter.

The best way to find Ursa Minor is to use Merak and Dubhe, the two 'Pointers' in Ursa Major, as direction-finders. An imaginary line from Merak, passing through Dubhe and prolonged, will come straight to Polaris, which is of magnitude 2.0 and lies within a degree of the polar point. It is highly luminous, and is the equal of 6000 Suns, but it is 680 light-years away.

The apparent distance between Polaris and Dubhe is about five times that between the Pointers; that is to say, between 25 and 30 degrees of arc.

The rest of Ursa Minor is easily located as soon as Polaris has been found. It resembles a dim and distorted Ursa Major, stretching down in the general direction of Mizar, and ending in two stars which are bright enough to be conspicuous – Beta, or Kocab (2.1), and Gamma (3.0). Kocab, sometimes called 'the Guardian of the Pole', is decidedly orange. The real luminosity is a hundred times that of the Sun.

The rest of the stars in the Ursa Minor pattern are fainter, so that mist or moonlight will drown them. When the sky is dark and clear, however, they are obvious enough, and in spite of its dimness the outline of Ursa Minor is unmistakable.

CASSIOPEIA

Cassiopeia, the proud queen of the Perseus legend, is marked by five stars making up a prominent W or M shape. A line from Mizar through Polaris, extending for about the same distance on the far side, will lead straight to Cassiopeia, so that in effect Cassiopeia, Polaris and the handle of the Plough or Big Dipper are lined up. When Ursa Major is high, Cassiopeia is low, and vice versa.

The stars of the W are Alpha or Shedir (magnitude 2.2), Beta (2.3), Gamma (variable; usually around 2.2), Delta (2.7) and Epsilon (3.3). Alpha, which is orange in colour, is a suspected variable, though with a very small range. Gamma is definitely variable, and is an unstable star. Usually it is about equal to Beta, but since I began following it, in 1935, it has ranged between magnitudes 1.6 and 3.3. It suffers occasional outbursts, and can sometimes outshine the Pole Star. Since it is always capable of springing surprises, it is worth observing. Beta makes an excellent comparison star.

Another unusual variable is Rho Cassiopeiae, close to Beta. Usually it is of around magnitude 5, but at irregular intervals it falls to below 6, so that it is then invisible with the naked eye – though this has not happened now for over forty years. Suitable comparison stars are Tau (5.1) and Sigma (4.9). Rho is inconveniently faint to be studied with the naked eye, but it can be done, though personally I always use wide-field binoculars.

The Milky Way flows through Cassiopeia, and there are some rich star-fields in the area.

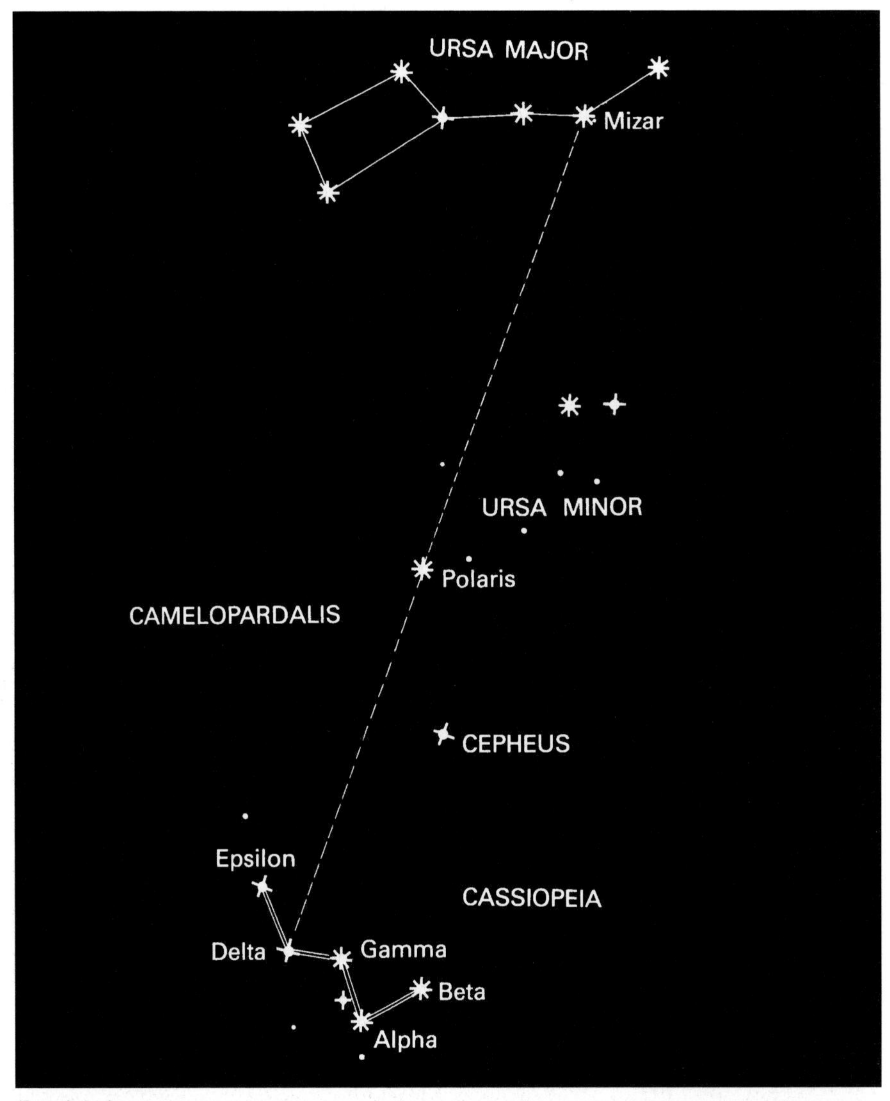

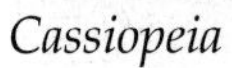
Cassiopeia

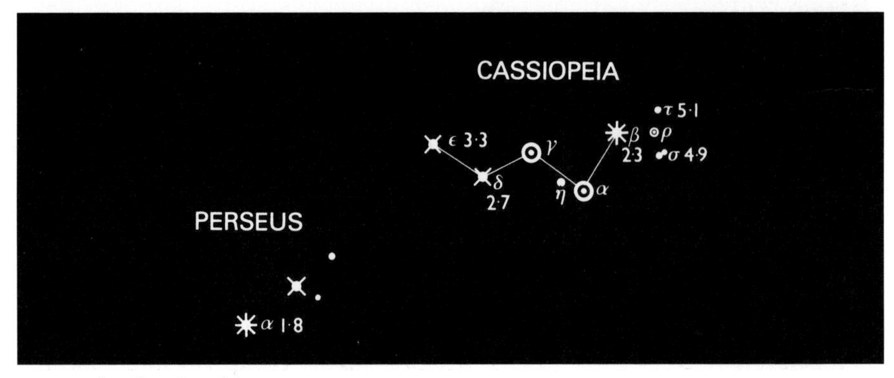

γ, α *and* ρ *Cassiopeiae*

CEPHEUS

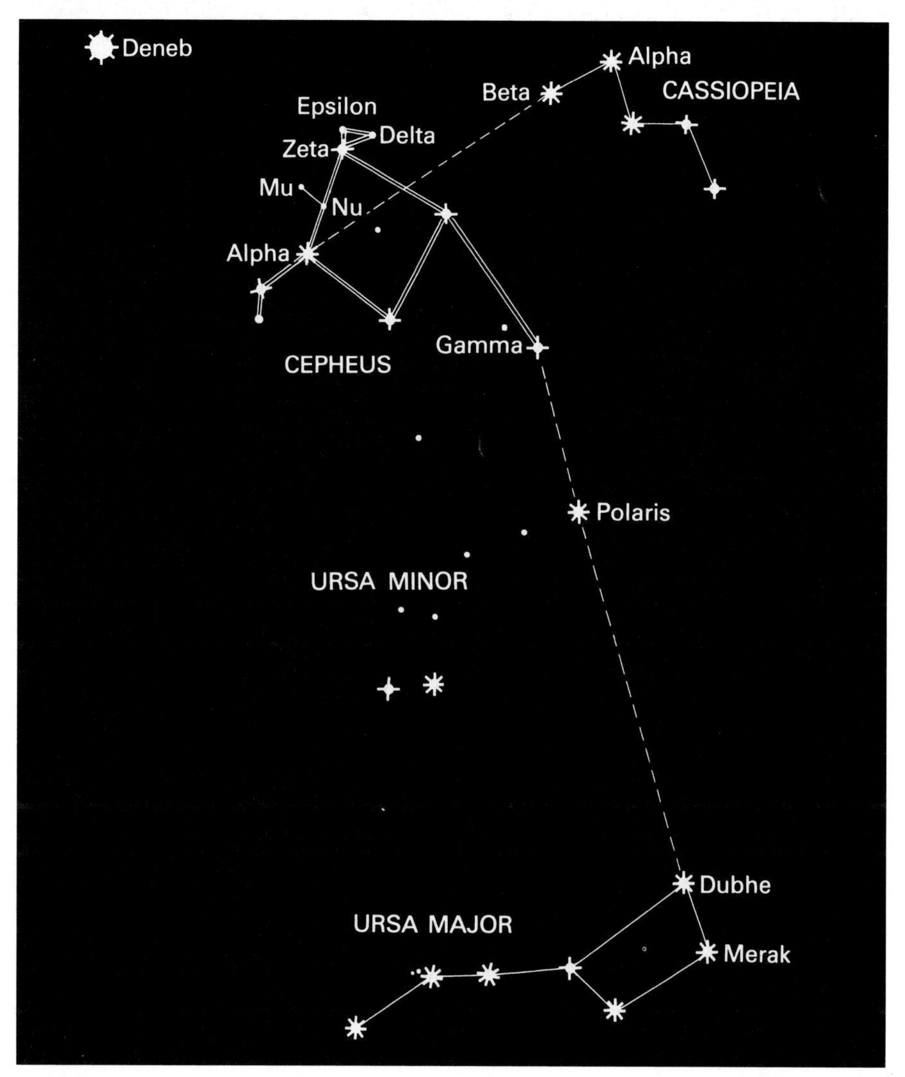

Cepheus

It has been said that 'the female of the species is more deadly than the male', and certainly King Cepheus is much less obtrusive than his wife. In fact he is not particularly easy to identify, since he is rather faint and has no obvious shape.

I have found that there are two good methods. The first is to continue the line from Merak and Dubhe beyond Polaris; the line will reach Gamma Cephei (3.2). The second guide is given by Cassiopeia, since two of the W stars (Alpha and Beta) show the way to Alpha Cephei (2.4). The whole of Cepheus lies inside the triangle bounded by Polaris, Cassiopeia and the first-magnitude star Deneb in Cygnus, which is just circumpolar in Britain but which sets briefly from New York.

Cepheus takes the form of a fairly large, rather faint diamond of stars, of which Alpha is the leading member. Yet vague though it may look, it contains two objects of exceptional interest. These are the variable stars Delta and Mu.

I have already said something about Delta, which has given its name to a whole class of vitally important variables – the Cepheids. It is one of a small triangle of stars lying between Cassiopeia and Deneb; the other two are Zeta (3.3) and Epsilon (4.2). The period of variation is 5.37 days, so that anyone with a little patience can check its fluctuations from night to night.

The second variable is Mu, about midway between Alpha and the small triangle which includes Delta. Mu Cephei has the distinction of being the reddest star visible without a telescope; the great observer Sir William Herschel called it 'the Garnet Star'. The light-changes are irregular, and the range is from magnitude 3.6 to well below 5. Nearby there is a useful comparison star, Nu (4.5), which looks dim enough but is in fact very remote, and is about 70,000 times as powerful as the Sun.

Really keen-sighted people claim that they can detect the redness of Mu Cephei with the naked eye, but the light intensity is too low for the colour to be conspicuous. Use binoculars, and the transformation is immediate: Mu looks like a glowing coal. It is at least 50,000 Sun-power, and so is much more luminous than Betelgeux in Orion, but of course it is also much further away.

Cepheus is so near the pole that from Britain or New York it is always fairly high up. Though it is a relatively barren group, the presence of the two fascinating variables redeems it.

LACERTA: *the Lizard*

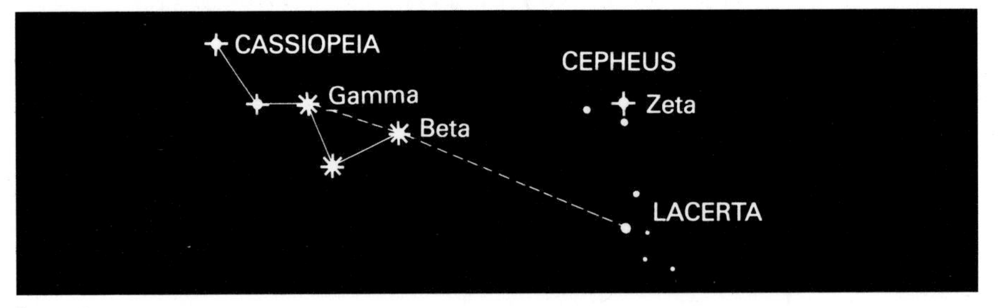

Lacerta

Adjoining Cepheus and Cassiopeia is a much smaller group, Lacerta, introduced by Hevelius of Danzig (now Gdańsk) in 1690. Hevelius was an energetic and skilful observer, and his star catalogue was good for its time. Altogether he drew up eleven new constellations, of which nine are still to be found on modern maps while the other two have been tacitly forgotten.

Frankly, there was little reason for making Lacerta into a separate group. One way to find it is to use the W of Cassiopeia, since Gamma and Beta point directly to it; but there are no stars in it brighter than magnitude 3.8, and neither are there any interesting objects. During autumn evenings it passes practically overhead.

DRACO: *the Dragon*

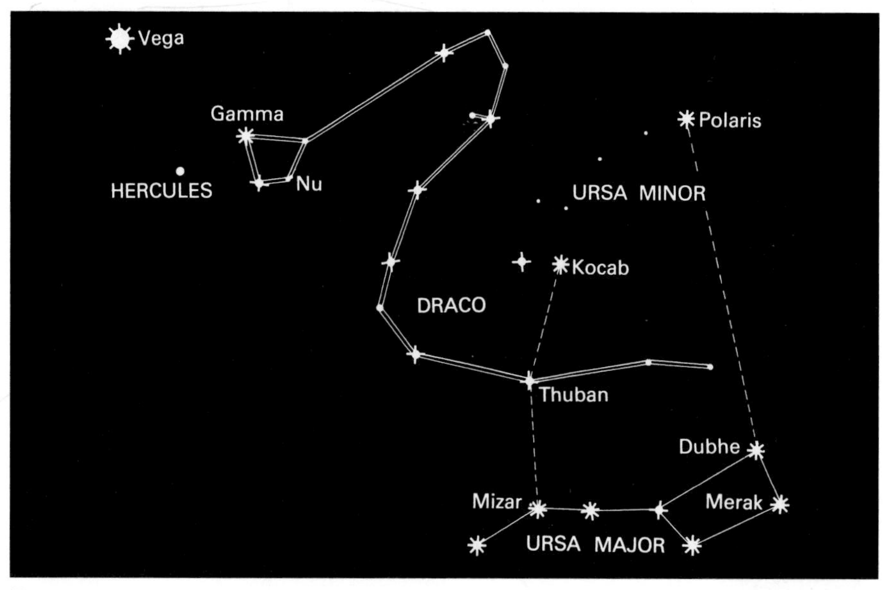

Draco

Our next constellation is the famous Dragon, which winds its way along in a line of stars not far from Ursa Major and Ursa Minor. Before coming to Draco, however, it will be helpful to make a slight digression and introduce two particularly brilliant stars, Capella and Vega, both of which are circumpolar in England and practically so from New York. Capella is the leader of Auriga (the Charioteer), while Vega is in Lyra (the Lyre or Harp).

Vega is of magnitude 0.0 and Capella 0.1, so that there is almost no difference between them; of all the stars visible from Europe, only Sirius and Arcturus are brighter. It so happens that they lie on opposite sides of the Pole Star, so that when Vega is high Capella is low. During summer evenings the lovely blue Vega is almost overhead, with Capella skirting the northern horizon; during winter evenings the position is reversed. If you see a bright star right above you, assuming that you are observing from Europe or North America, you may be quite sure that you are looking at either Capella or Vega.

In Star Map 1, the whole area containing these two stars has been shown. Capella is the easier to find, because Megrez and Dubhe in Ursa Major show the way to it. To identify Vega, either use Capella and Polaris as pointers or else work from Ursa Major, which means the awkward procedure of starting at Phad and taking a line midway between Megrez and Alioth.

Now we can return to Draco, which has no really bright stars. The stream begins roughly between Dubhe and Polaris, and winds its way between Ursa Major and Ursa Minor, making off in the general direction of Cepheus. It then turns, and ends at the Dragon's head, not far from Vega. In the head lies Gamma (2.2), ironically the brightest star in the whole constellation.

Look carefully at another of the stars in the head, Nu Draconis, and you will see that it is double. Keen-sighted people can split it easily, though I admit that I find it difficult; the components are equal at magnitude 4.9. The twins of Nu Draconis are genuinely associated, but are very wide apart.

Alpha Draconis, which lies between Mizar and Kocab, used to be the pole star in ancient times, but its magnitude is only 3.6. It is still often known by its old proper name, Thuban.

CAMELOPARDALIS: *the Giraffe*

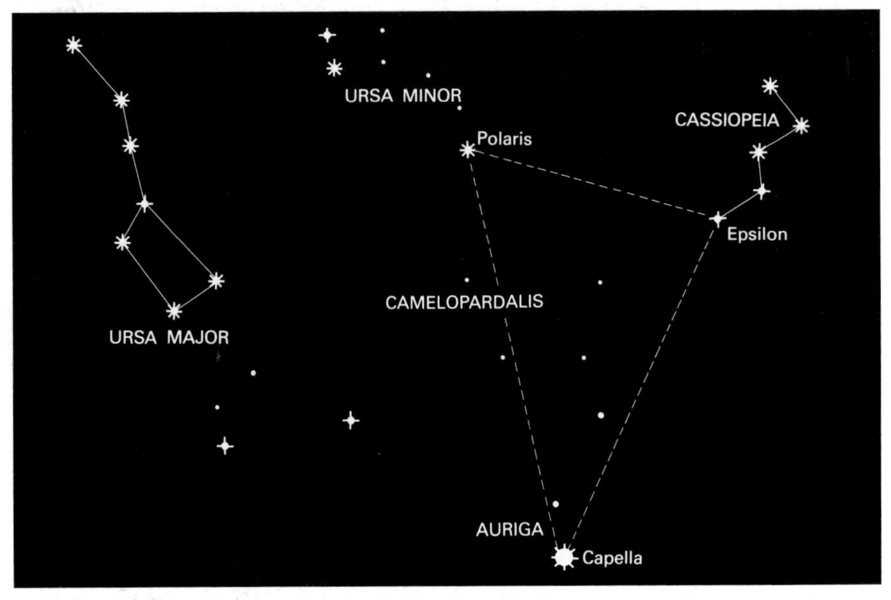

Camelopardalis

Camelopardalis is without doubt one of the dullest constellations in the sky, with no bright stars or objects of interest. It fills the triangular area bounded by Capella, Polaris and Epsilon Cassiopeiae in the W. Very little can be made out; there are a few stars of between the fourth and fifth magnitudes, but even these make up no definite pattern. It has been said that if you come across an area of the sky with nothing in it, you will have located the Giraffe.

LYNX: *the Lynx*

Next to Camelopardalis is the Lynx. It lies in the region between Capella to the one side and the Pointers, Merak and Dubhe, to the other. It also extends further south; not all of it is circumpolar in Britain, since a small part of it drops below the horizon. The Lynx, unlike the Giraffe, does contain one moderately bright star, Alpha (3.1), and there is a dim, inconspicuous line of stars extending beyond the end of Ursa Major, but on the whole the Lynx is hardly worth bothering about.

Lynx

CANES VENATICI: *the Hunting Dogs*

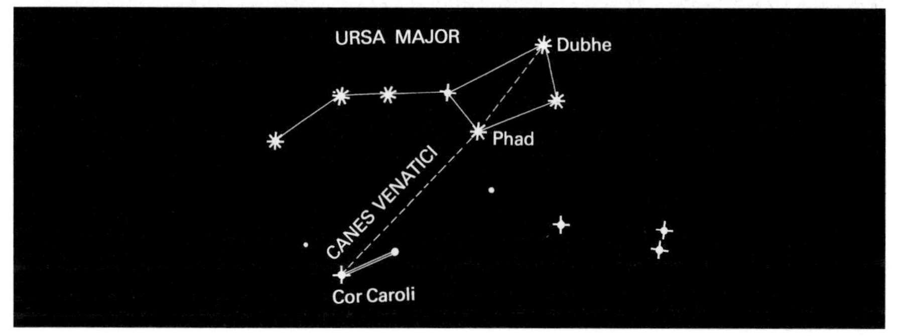

Canes Venatici

Finally in the list of northern circumpolar constellations we come to Canes Venatici, added to the sky by the first Astronomer Royal, John Flamsteed. The only bright star, Alpha (2.9), has been named Cor Caroli, or Charles's Heart, in honour of the murdered monarch Charles I. It lies in a rather isolated position not far from the tail of Ursa Major; Dubhe and Phad point roughly towards it. There are no other stars in the constellation above the fourth magnitude. Originally the maps showed two dogs, Asterion and Chara, being held on a leash by the herdsman Boötes. Canes Venatici never sets over Britain, but it can become very low, and from New York it dips below the horizon.

A NOTE ON THE STAR MAPS

The maps which follow for the northern hemisphere cover the whole sky as seen from Britain and New York. To select the correct monthly charts, refer to the table on p. 201.

7

Northern stars: the January sky

Monthly star charts are published in various journals, including a few national newspapers. Most of them are drawn in circular form, which is perfectly accurate but which makes the charts difficult to use. I have preferred to give two hemispherical maps for each month, one showing the northern aspect of the sky and the other showing the southern aspect. Unfortunately the constellations near the overhead point are badly distorted, and this would be a serious drawback for precision charts, but what I intend to do is to give maps which are best suited to star recognition. I have also refrained from putting in too many stars, so that the main patterns stand out clearly.

Each pair of charts has been drawn for 22 hours G.M.T. (Greenwich Mean Time). In astronomy the 24-hour clock is always used, and artificial manoeuvres such as British Summer Time or Daylight Saving Time are shunned. Therefore 22 hours G.M.T. is the same as 10 p.m., or 11 p.m. when Summer Time is in operation.

One month means a difference in star positions of approximately two hours, so that the set of charts may be adapted for any season. Thus the aspect at 10 p.m. on 1 January will be the same as that at 8 p.m. on 1 February, or 6 p.m. on 1 March. In fact, all you need do is subtract two hours a month. Working backwards through the year, you have to add two hours a month; thus the view at midnight on 1 December will be the same as that at 10 p.m. on 1 January.

The January charts shown here are for 22 hours G.M.T., or 10 p.m. The view will be the same on:

1 October:	4 a.m.	(4 hours G.M.T.)
1 November:	2 a.m.	(2 hours G.M.T.)
1 December:	midnight	(0 hours G.M.T.)
1 January:	10 p.m.	(22 hours G.M.T.)
1 February:	8 p.m.	(20 hours G.M.T.)
1 March:	6 p.m.	(18 hours G.M.T.)

So if you want to find out which groups are on view in the early hours of an October morning, simply look at the charts given here for January evening. The two-hour rule is not exact, but it is quite sufficient for most purposes.

It follows, of course, that what is usually called a 'winter constellation', such as Orion, is not visible only during the winter; it is seen just as well on an autumn morning. However, most people do their sky-watching before going to bed, so that charts drawn for evening aspects are more widely useful.

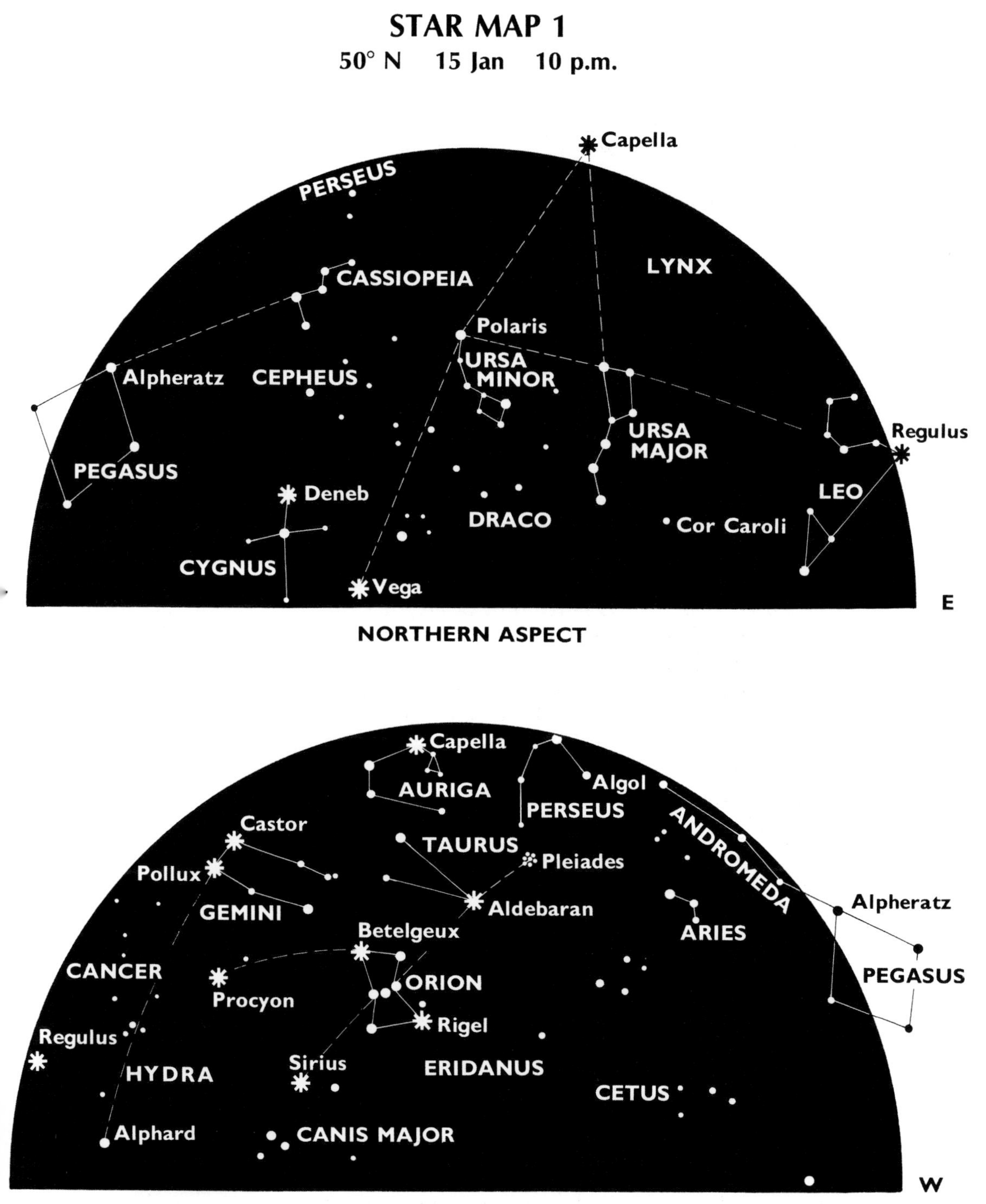
STAR MAP 1
50° N 15 Jan 10 p.m.
Capella
PERSEUS
CASSIOPEIA
LYNX
Polaris
URSA MINOR
Alpheratz
CEPHEUS
URSA MAJOR
Regulus
PEGASUS
Deneb
LEO
DRACO
Cor Caroli
CYGNUS
Vega
E
NORTHERN ASPECT
Capella
Algol
AURIGA
PERSEUS
ANDROMEDA
Castor
TAURUS
Pleiades
Pollux
Aldebaran
Alpheratz
GEMINI
ARIES
Betelgeux
CANCER
PEGASUS
ORION
Procyon
Rigel
Regulus
Sirius
ERIDANUS
HYDRA
CETUS
Alphard
CANIS MAJOR
W
SOUTHERN ASPECT

If you look north-east late on a January evening, the first group you will notice will certainly be Ursa Major, with its tail pointing towards the horizon. Polaris, of course, will be in its customary position, due north (it can never be anywhere else), and Cassiopeia will be high in the north-west. There should be no trouble in finding Cepheus, Draco and the other circumpolar groups, though you may lose the Dragon's head in horizon mist.

Capella is practically at the zenith, or overhead position. It belongs properly on the southern chart, but I have shown it here also so that it can act as a link between the two. Since Capella is at its best, Vega, on the opposite side of the pole, is as low down as it can ever be; it lies close to the British horizon, and from New York it cannot be seen at all. Rather higher than Vega is the bright star Deneb, leader of Cygnus (the Swan).

If you use the Pointers, Dubhe and Merak in Ursa Major, in the reverse direction – that is to say, away from Polaris – the line will reach the large, bright group of Leo, the Lion, which is rising in the east and will be very prominent during evenings in spring. Over in the west, the Square of Pegasus is dropping towards the horizon.

Pegasus is classed as an autumn constellation, and I will describe it with the October charts; for the moment, suffice to say that you can find it by using two of the stars in the W of Cassiopeia as direction-finders.

Extending between Pegasus and the equally bright group of Perseus is a line of bright stars marking Andromeda. In point of fact Andromeda connects Pegasus with Perseus, but it too is probably best classed with the autumn groups even though it is still high up during early evenings in January. The almost blank area above Ursa Major indicates the large, dim constellations of Camelopardalis and Lynx.

The southern aspect is dominated by Orion, which cannot possibly be overlooked. Its stars are much brighter than those of Ursa Major, and two of them, Betelgeux and Rigel, are well above the first magnitude. Since all Orion's neighbours will be described in the following pages, there is no point in saying more about them in a general account of the January view – except to add that Sirius, the Dog-Star, is now at its best, and twinkles gloriously above the southern horizon.

ORION

Orion is possibly the most magnificent constellation in the entire sky. The main outline is formed by seven stars, which are often known by their proper names as well as their Greek letters and which are extremely useful as direction-indicators. They are:

Beta or Rigel (magnitude 0.1)
Alpha or Betelgeux (variable between 0.2 and 0.8)
Gamma or Bellatrix (1.6)
Epsilon or Alnilam (1.7)
Zeta or Alnitak (1.8)

Kappa or Saiph (2.1)
Delta or Mintaka (2.2)

Rigel is a true celestial searchlight. It is so remote that its light takes around 900 years to reach us, so that we are now seeing it as it used to be in the time of William the Conqueror. Yet even at this tremendous distance Rigel still appears as the seventh brightest star in the sky; its true luminosity is 60,000 times that of the Sun.

It is hard to appreciate the amount of energy being radiated by this titanic star. The Sun is a glow-worm by comparison; were Rigel as close as, say, Sirius it would be visible in broad daylight, and even at 900 light-years – well over 5000 million million miles – it is still striking. It is pure white, and because it is so energetic its lifetime will be much shorter than that of our Sun.

Betelgeux is completely different; it is a huge red supergiant with a diameter of 250 million miles, large enough to swallow up the whole orbit of the Earth round the Sun. Like so many red stars, Betelgeux is variable, with a very rough period of about five years. Sometimes it almost equals Rigel, while at others it is little brighter than Aldebaran in the Bull.

The only really useful comparison stars are Capella (0.1), Rigel (also 0.1), Procyon (0.4) and Aldebaran (0.8), but there is the problem of altitude; for example, Rigel is always lower than Betelgeux as seen from the northern hemisphere of the Earth, so that allowance has to be made for extinction. Moreover, the changes in Betelgeux are slow. Its luminosity is some 15,000 times that of the Sun – powerful indeed, though by no means comparable with Rigel.

Incidentally, the name Betelgeux may be spelled in various ways; Betelgeuse and Betelgeuze are other versions. Arab scholars tell me that it really means 'shoulder', and is best pronounced 'bettle-gurz', though many people refer to it as 'beetle-juice'!

The other bright stars of Orion are hot and white, similar in type to Rigel and, in the case of Saiph, almost as luminous. Alnilam, Alnitak and Mintaka make up the famous Belt, which is so useful as a direction-marker. Mintaka is very slightly variable, but I doubt whether its changes are detectable with the naked eye.

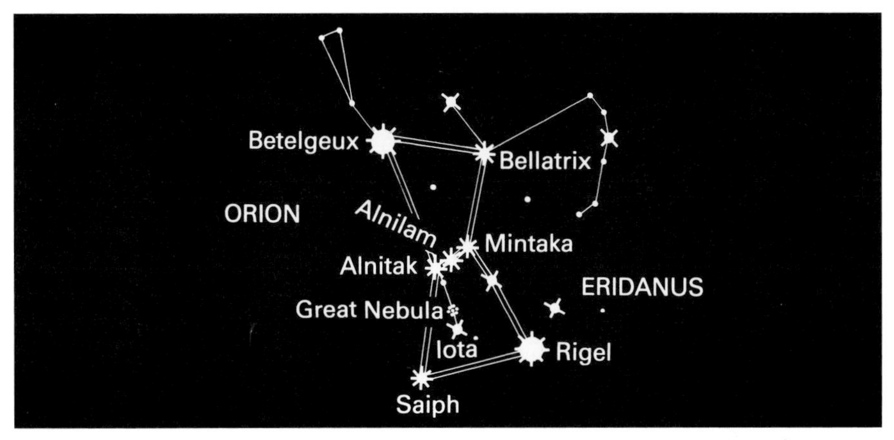

Orion

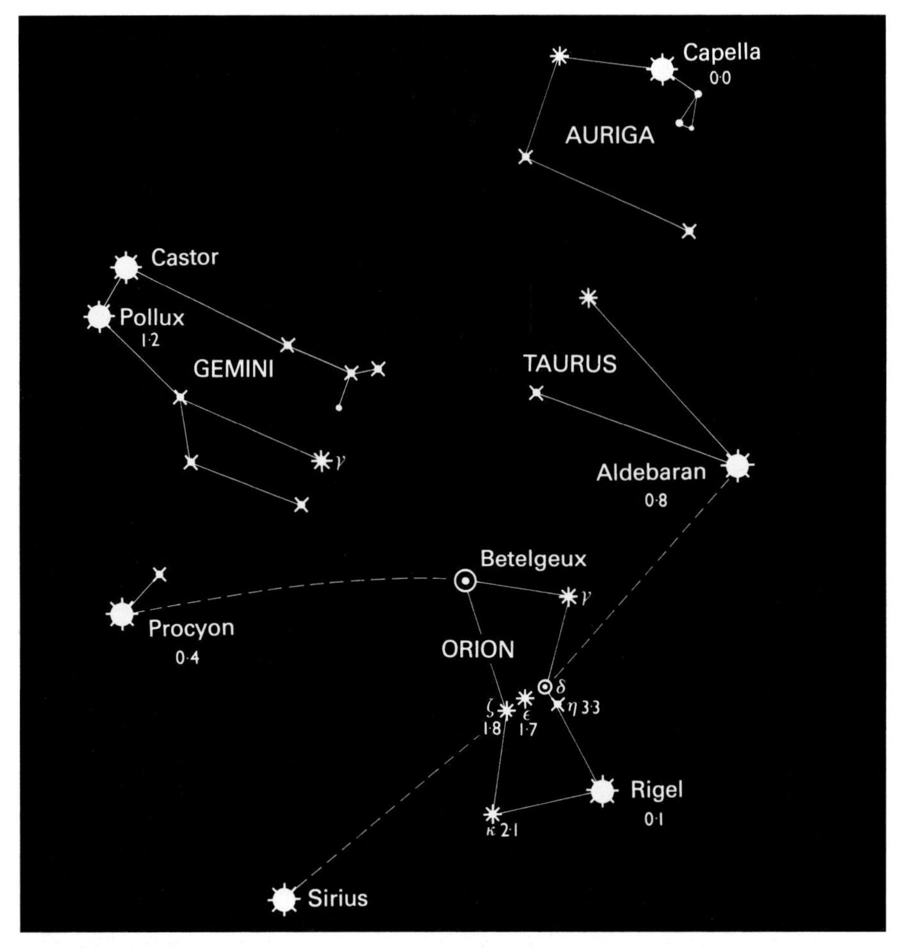

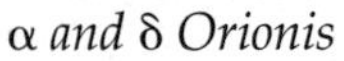
α and δ Orionis

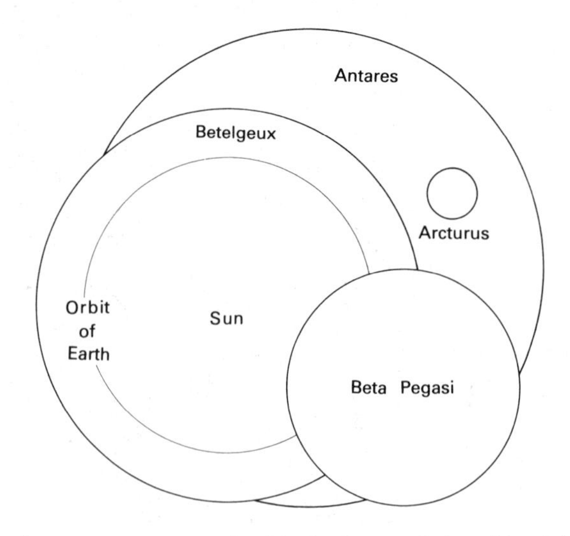

Sizes of some large stars, compared with the Sun and the orbit of the Earth.

The Sword of Orion extends southward from the Belt. It contains yet another remote, highly luminous white star, Iota (2.8), but the most notable feature of the Sword is the misty patch of the Great Nebula, M. 42. On a clear night the Nebula is easily visible with the naked eye, and binoculars show it in its true guise as a patch of shining gas. It is probably the best known of all gaseous nebulae; deep inside it is a mysterious object which we can never see, but which we can detect because of its long-wavelength (infra-red) radiation. M. 42 is a stellar birthplace, and contains many very young stars which have not yet settled down and become stable.

The Nebula is over 1000 light-years away. Near its edge, on the side facing us, is the multiple star Theta Orionis. This has four main components arranged in the form of a trapezium, and it is these which make the Nebula shine. A small telescope will show them, but with the naked eye they appear only as a dim speck.

Orion covers a wide area of the sky, and spreads out well beyond the brilliant seven-star pattern. Note, for instance, the long line of fainter stars extending in a roughly north–south direction some way from Bellatrix. As a guide to other groups, Orion is unrivalled. For part of the year it is invisible, because it is too near the Sun; June, July and August skies know no Orion, and are the poorer for it, but in winter the Hunter dominates the night scene, and remains visible until disappearing into the evening twilight during springtime.

CANIS MAJOR: *the Great Dog*

During evenings in January, Sirius shines brilliantly in the south. It is a pure white star, but the fact that it is so bright, and always rather low down,

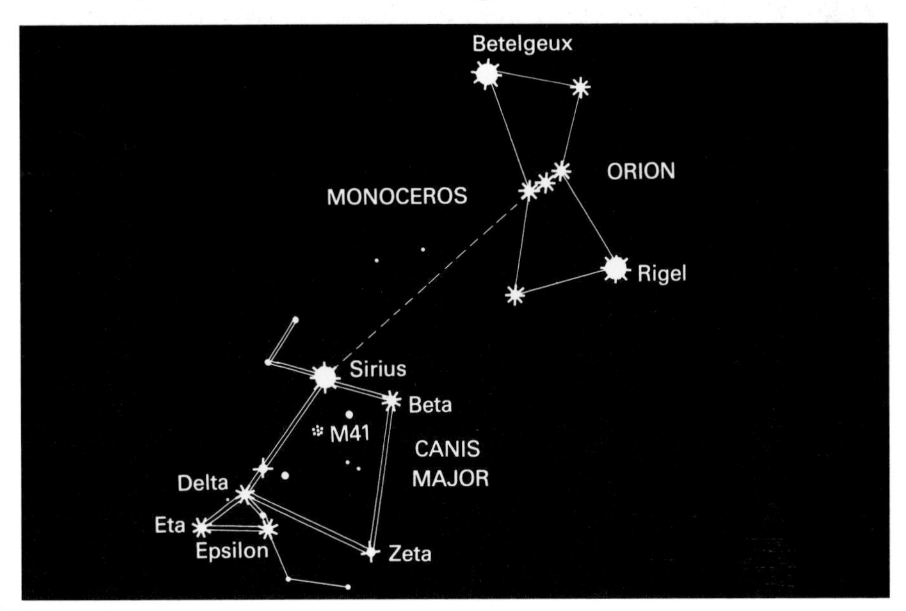

Canis Major

makes it twinkle violently, flashing various colours. It lies in line with the three stars of Orion's Belt, but no pointers to it are really needed. Yet it is not exceptionally luminous; it is 26 times as powerful as the Sun – puny indeed compared with Rigel – but it is one of our closest stellar neighbours, and is only 8½ light-years away from us. Seen through binoculars or a telescope, the many-hued flashing of Sirius is a lovely sight.

Canis Major contains several other bright stars, though all are completely outshone by Sirius. Beta (2.0) lies west of Sirius; well south of them is a triangle made up of Delta (1.9), Eta (2.4) and Epsilon (1.5), never conspicuous from Britain but easily seen in New York and very prominent from southern Europe or the southern United States. All three are very luminous and remote. Delta, indeed, is probably well over 120,000 times as powerful as the Sun, but it is a full 3000 light-years away.

There is an interesting star-cluster in Canis Major, near Sirius and Beta, known as M. 41. It is just visible with the naked eye on a clear night. I always find it difficult to locate from England because of its low altitude, but from southern latitudes it is easy enough to find. It is an open cluster, of the same basic type as the Pleiades.

The Milky Way runs through a part of Canis Major, but misses Sirius by a fair margin.

CANIS MINOR: *the Little Dog*

Canis Minor, the second of Orion's two dogs, is marked by the bright star Procyon; its magnitude is 0.4, not greatly inferior to Rigel. The best way to find it is to use Bellatrix and Betelgeux as pointers. Extend a line from one to the other, and continue it for some distance from Betelgeux, curving it slightly. Procyon will be found at once, partly because it is so bright and partly because there are no comparable stars anywhere near it. On January evenings it is high up, and is visible for most of the night, since it does not set until about six o'clock in the morning.

Procyon, like Sirius, is a near neighbour, 11½ light-years away and eleven times as luminous as the Sun. It is a pure white star as seen with the naked eye, though its surface is cooler than that of Sirius. The only other moderately bright star in Canis Minor is Beta (2.9).

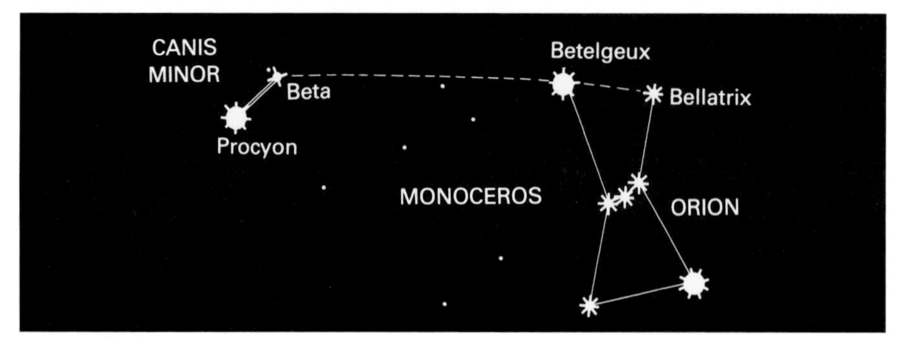

Canis Minor

MONOCEROS: *the Unicorn*

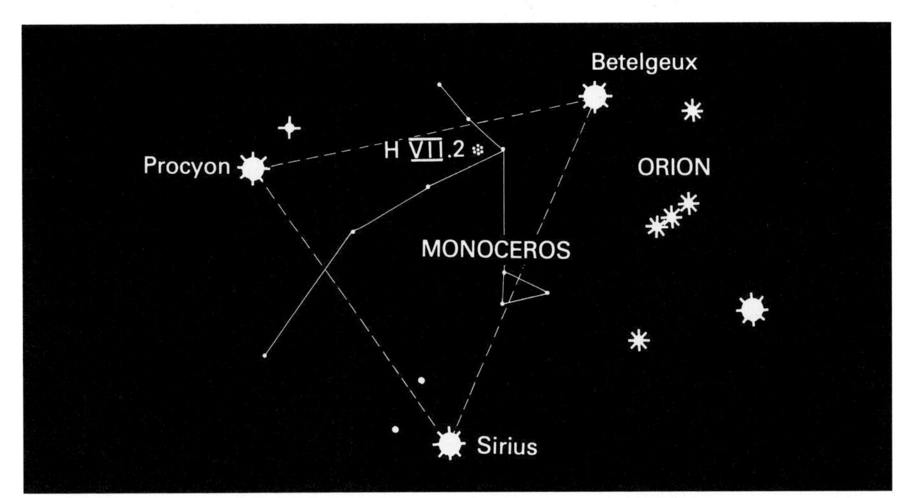

Monoceros

Monoceros, one of the constellations named by Hevelius in 1690, is nothing like a horned horse; there are no bright stars, and no distinctive pattern. However, it is easy to find because it lies inside the large triangle formed by Betelgeux, Procyon and Sirius.

Actually, Monoceros is not as dull as might be thought. The Milky Way flows through it, and the whole region is very rich in faint stars. Also, there are various clusters, of which one – roughly between Procyon and Betelgeux – is just visible with the naked eye. It was not listed by Messier, and is known either as 12 Monocerotis, or as H. VII. 2, the second object in a later catalogue drawn up by Sir William Herschel. It is C.50 in the Caldwell catalogue. It is not easy to identify because there are so many other faint stars nearby, and it is not of much interest to the naked-eye observer.

PUPPIS: *the Poop*

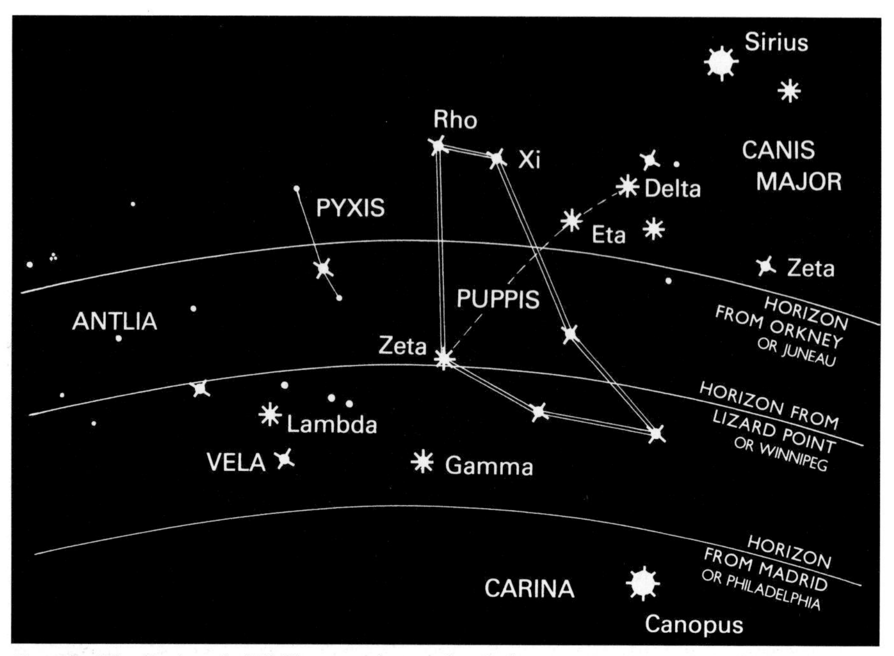

Puppis. The limits of visibility are shown for Orkney or Juneau, Lizard Point or Winnipeg, and Madrid or Philadelphia.

In older maps the great constellation of Argo Navis, the Ship Argo, covers a huge area. Today it has been divided up. Most of it is invisible from Europe and much of North America, but a part of the poop, Puppis, does rise above the horizon. Rho (2.8) lies some way east of the triangle formed by Delta, Epsilon and Eta Canis Majoris, with Xi (3.5) nearby. New Yorkers, though not Londoners, may catch sight of Zeta (2.2), which is another cosmic searchlight, fully as luminous as Rigel. Zeta Puppis may be found by using Delta and Eta Canis Majoris as pointers, but nobody who lives north of latitude 50° N will be able to see it. The Milky Way flows through Puppis, and continues down over the southern horizon.

Zeta does theoretically rise from Lizard Point, the southernmost tip of England, but I have never been able to glimpse it.

PYXIS: *the Compass* and **ANTLIA:** *the Air-Pump*

Adjoining Puppis are two utterly insignificant constellations, Pyxis (the Argo's old compass) and Antlia. Since neither contains anything of interest, and there are no stars above the fourth magnitude, it seems pointless to provide separate maps for them, and I have simply shown their positions on the chart for Puppis. From Britain, at least, it is doubtful whether any of the stars of Antlia will be visible with the naked eye, because they are too dim and too low down. I have looked for them on more than one occasion, but without success.

GEMINI: *the Twins*

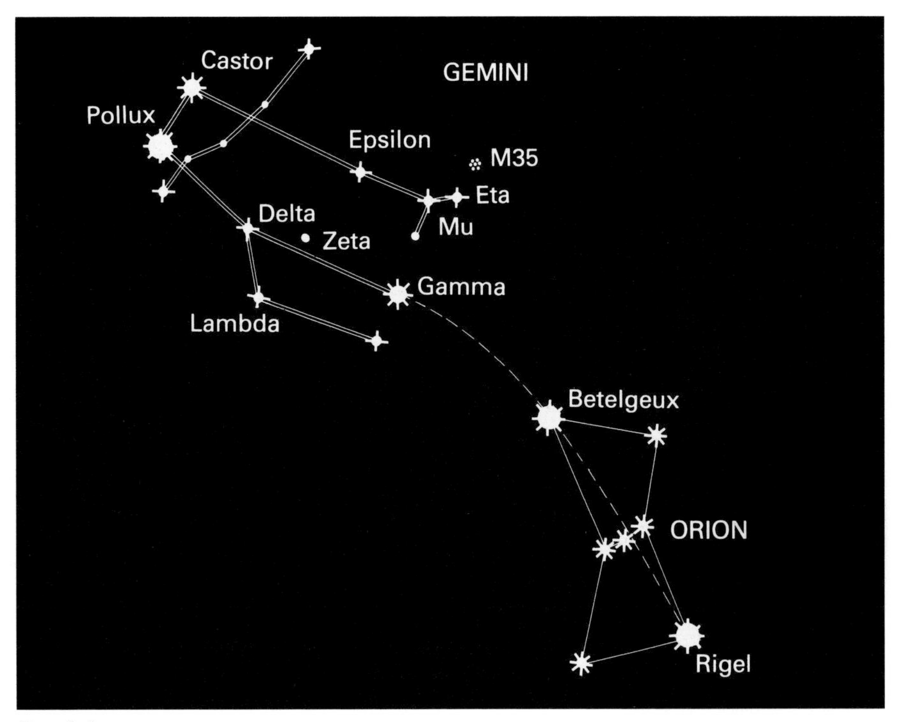

Gemini

From these faint, elusive groups we turn next to a really brilliant constellation, Gemini, which lies in the Zodiac and may therefore contain planets. The two leading stars, Castor and Pollux, are named in honour of the twin boys who accompanied Jason in the Argo during the quest for the Golden Fleece. The stars are only 4½ degrees of arc apart, so that they are rather closer together than the pointers in Ursa Major. To find them, draw a line from Rigel through Betelgeux and extend it for roughly double the distance across Orion. The Twins lie well north of Orion in the sky, and are often visible when Orion is not; by six o'clock on a January morning, for instance, the Hunter has set but the Twins are still on view. Luckily, Castor and Pollux may also be found by using our other main 'anchor', Ursa Major. The direction-finders in this case are Megrez and Merak; the region between Merak and the Twins is occupied by the outer parts of Ursa Major and the faint, dull Lynx, so that the line will pass near no bright stars until it reaches Castor and Pollux.

Pollux is the brighter of the two; its magnitude is 1.1, as against 1.6 for Castor, so that the difference is quite noticeable. Moreover Pollux is obviously orange, whereas Castor (a multiple system, as we have noted) is white. They are not true neighbours; Castor is 46 light-years away from us, Pollux only 36.

The rest of Gemini is made up of long streams of stars, stretching from the Twins in the general direction of Orion. Gamma, between Betelgeux on the one side and Castor and Pollux on the other, is of magnitude 1.9, and is bright enough to be conspicuous. Near it are two less prominent stars, both of which are of interest. Mu (2.9) is a Red Giant, though its colour is not obvious with the naked eye. Its neighbour, Eta, is also a Red Giant; it is variable, with a magnitude range of from 3 to 4. The period is said to be around 231 days, but I am rather sceptical about this, and generally the magnitude is between 3¼ and 3½.

Zeta Geminorum is another variable, this time a Cepheid. The range is from 3.7 to 4.3 so it is not brilliant, but it is conveniently placed between Gamma and Pollux, slightly closer to Gamma. The period is 10.1 days, and since this is longer than that of Delta Cephei (5.3 days), Zeta Geminorum is the more luminous of the two. At its best it is not much fainter than its neighbours, Delta and Lambda.

The Milky Way runs right through Gemini, and the whole region is very rich. Look for the dim patch near Eta. This is in fact a lovely open cluster, M. 35. Binoculars show it well.

AURIGA: *the Charioteer*

On a January evening the overhead position is occupied by Capella, the sixth brightest star in the sky and only fractionally below magnitude zero. It can be found by using Ursa Major, and we can also use Orion, since a line drawn from Mintaka through Bellatrix leads almost straight to it.

Capella, like the Sun, is yellow, but it has 160 times the Sun's power. Actually it is a binary, but the two components are much too close together to be seen separately except with giant telescopes. It is circumpolar from Britain, though on summer evenings it almost touches the southern horizon, and from New York it sets briefly. It is interesting to compare it with Vega, on the opposite side of the pole. The two are so nearly equal in magnitude that in general the star which is the higher up will appear the brighter of the two.

Auriga contains several bright stars in addition to Capella: Beta (1.9), Theta (2.6) and Iota (2.7) complete a sort of kite-shape. There is also a bright white star of magnitude 1.6, with a proper name of Al Nath, lying between Capella and Betelgeux. It used to be included in the Charioteer, as Gamma Aurigae, but for some reason or other it has been transferred to the neighbouring constellation of the Bull, as Beta Tauri. There are grounds for suggesting that it would have been better left with Auriga, since it fits well into the kite-pattern, whereas Taurus has no definite shape at all.

Near Capella you will see a triangle of fainter stars, nicknamed the Haedi or Kids. They look unspectacular, but two of them are remarkable objects. Epsilon Aurigae, at the apex of the triangle, is a binary of most unusual type. The brighter star is an exceptionally powerful yellow supergiant, at least 200,000 times as luminous as the Sun, and over 4500 light-years away. The fainter component has never been seen, and we are not even sure whether it

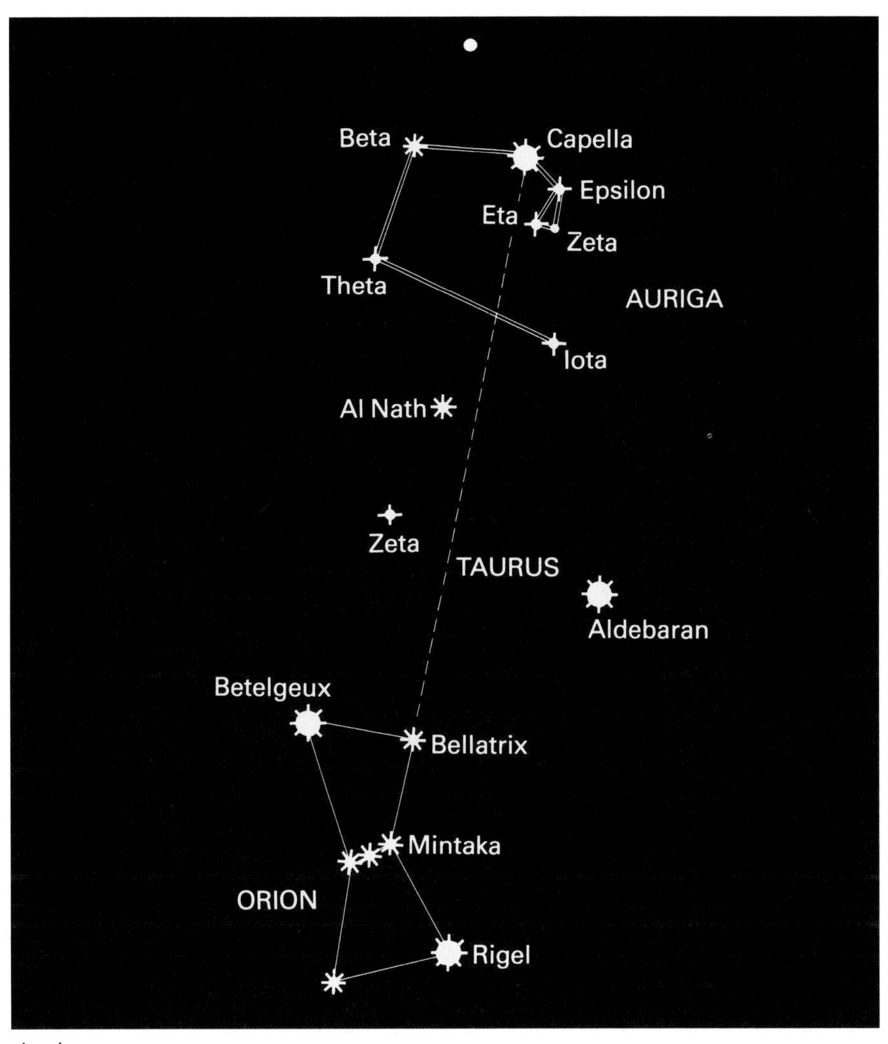

Auriga

is a normal star. Every twenty-seven years it passes in front of the supergiant, and the total magnitude drops to below 3½ but as the last eclipse ended in the summer of 1984, nothing more will happen until 2011.

The dimmest member of the triangle, Zeta, is of the same kind, though the period is shorter (972 days) and the magnitude range is very small (3.7 to 4.2). It is pure coincidence that these two eclipsing binaries lie side by side in the sky; Zeta is much the nearer of the two.

Part of Auriga is circumpolar from Britain, but not from the United States (except Alaska). The Milky Way runs through it, so that it is extremely rich in faint stars.

PERSEUS

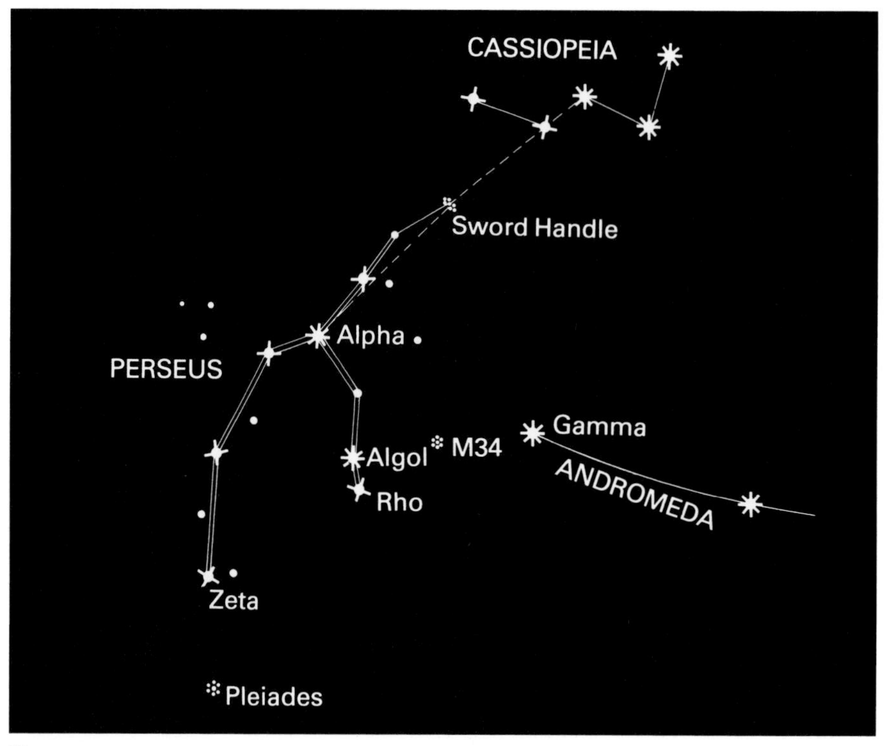

Perseus

Perseus, the gallant hero of the legend about the princess and the sea-monster, adjoins Auriga. It may be said to belong to both the winter and autumn charts. It must be described here because it is high up during January evenings, but in many ways it belongs more comfortably in the pattern associated with Andromeda and Pegasus. From Perseus, the line of stars marking Andromeda extends towards the western horizon.

So far as Perseus is concerned the best pointer is Cassiopeia, because two of the W stars – Gamma and Delta – show the way to it. Ursa Major is too far off to be used conveniently, and neither are there any obvious guides from Orion, but on a clear night there should be no problem; simply trace the path of the Milky Way from Capella to Cassiopeia and you are bound to locate Perseus. The group has a shape which is easy to recognize, and there is one star, Alpha, as bright as magnitude 1.8.

Of course the most famous star in the constellation is the eclipsing binary Algol. The usual magnitude is 2.4, but every 2 days 13 hours it fades, dropping to magnitude 3.3 and remaining at minimum for twenty minutes before starting to recover. As we have seen, it is not truly variable, but is periodically covered up (partially, at least) by its dimmer companion. The times of the minima can be predicted very accurately, and are listed in publications such as the *Handbook of the British Astronomical Association* or *Sky and Telescope* magazine. When a 'wink' is due, observations carried out at, say,

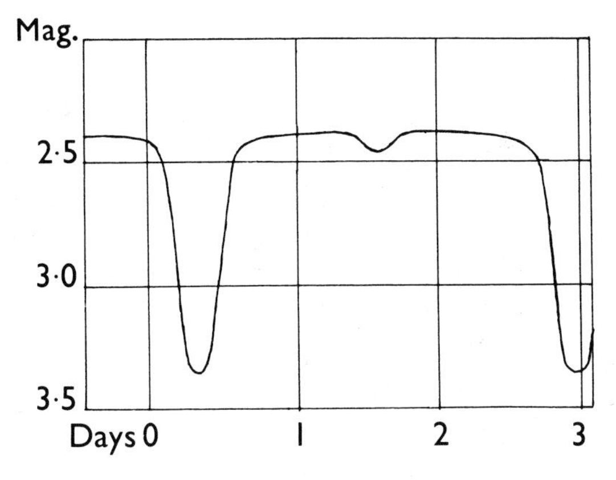

Light-curve of Algol

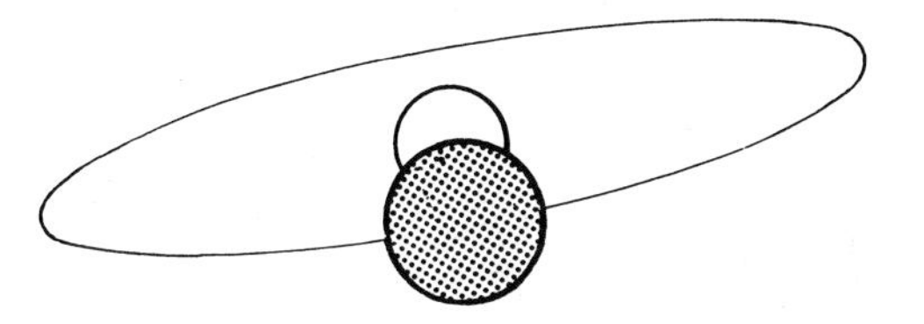

The Algol pair, with the larger, fainter component concealing as much of the brighter star as it ever can – so producing the chief minimum (mag. 3.3).

10-minute intervals will allow a light-curve to be drawn up. If you happen to look at Perseus and see that Algol is unexpectedly faint, you may be sure that an eclipse is in progress.

Next to Algol in the sky, but not in any way associated with it, is a genuine variable star, Rho Persei. It is a semi-regular Red Giant, fluctuating between magnitudes 3.25 and just below 4; there is said to be a rough period of about five or six weeks, but it is very rough. Kappa (3.8) is the best comparison star. The colour of Rho is obvious with binoculars, but I doubt whether many people will notice it with the naked eye.

Forming a triangle with Algol and Rho is an open star-cluster, M. 34, which is just visible without optical aid under good conditions. However, there is a much more interesting object: the Sword-Handle, Chi-h Persei (Caldwell-14) or H. VI. 33-4 (not to be confused with the nebula in the Sword of Orion). The Sword-Handle is easy to locate as a misty patch. Gamma and Delta Cassiopeiae, in the W, show the way to it; all you have to do is extend the line for about twice the distance between Gamma and Delta. The Sword-Handle is made up of two clusters close together; seen with any telescope it is a lovely sight, since the two clusters lie in the same low-power field.

The Milky Way is particularly bright in Perseus, and much of the constellation is covered. Note, too, the end star in the main pattern, Zeta (2.8). It looks commonplace enough, but it is in fact a very luminous White Giant well over 100 light-years away.

TAURUS: *the Bull*

Taurus is another splendid winter constellation. Here again there is no problem in finding it whenever Orion can be seen, because the Belt stars point upwards to Aldebaran, leader of Taurus. Aldebaran is an orange star of magnitude 0.8, and thus looks like a slightly less brilliant edition of Betelgeux. It is not appreciably variable.

Aldebaran is apparently a member of a scattered cluster of stars known as the Hyades. Several members of the cluster are visible with the naked eye, extending from Aldebaran in a sort of V-formation; most of them are between the third and fourth magnitudes, while one, Theta Tauri, is a double wide enough to be split without optical aid.

I say that Aldebaran is 'apparently' in the Hyades because it is not a genuine member of the cluster at all; it is seen in the foreground, so to speak. It is 68 light-years away from us, while the Hyades themselves are twice as remote. Aldebaran is as far from the Hyades as we are from Aldebaran.

The Pleiades, or Seven Sisters, also lie in Taurus. To find them, continue the line from Orion's Belt through Aldebaran, and curve it slightly. The Pleiades will be found at once; at first glance they look like a misty patch, but a closer look will show at least seven individual stars if the sky is clear and dark. One of them, Alcyone or Eta Tauri, is of magnitude 2.9.

As well as Alcyone, Electra and Atlas are well above the fourth magnitude. Then come Merope, Maia, Taygete, Celaeno, Pleione and Asterope. Binoculars show many more Pleiads, and the whole cluster contains several hundred stars, perhaps as many as 500 within a radius of 25 light-years. The distance from Earth is about 410 light-years. All the leaders are hot and white, and the cluster is young by cosmical standards.

It is interesting to decide how many individual stars can be made out. In a *Sky at Night* television programme over twenty years ago I invited viewers to look at the Pleiades under good conditions, and note the number of stars seen. The results were interesting. Hundreds of replies were received, and most of them showed seven stars. A few keen-eyed people recorded eight, nine or even (in one case) eleven, but the average worked out at seven, so the nickname really is appropriate.

Moonlight will hide all the Pleiades except Alcyone, and even Alcyone vanishes when the Moon is near full and in that part of the sky. Mist will also conceal the group; but against a dark background the Pleiades are quite unmistakable, and make up by far the most conspicuous of the open clusters.

Lambda Tauri, not far from the Hyades, is an Algol-type eclipsing binary. As its range is 3.2 to 4.2 it is always a naked-eye object, and there are some convenient comparison stars. The period is just under four days.

Zeta Tauri (3.0) lies between Al Nath and Betelgeux; it was near here that a brilliant supernova flared out in the year 1054, and has left the gas-patch of the Crab Nebula, which is unfortunately well below naked-eye visibility. Al Nath, as we have noted, is now included in Taurus rather than Auriga. Its magnitude is 1.6, almost exactly the same as that of Bellatrix in Orion.

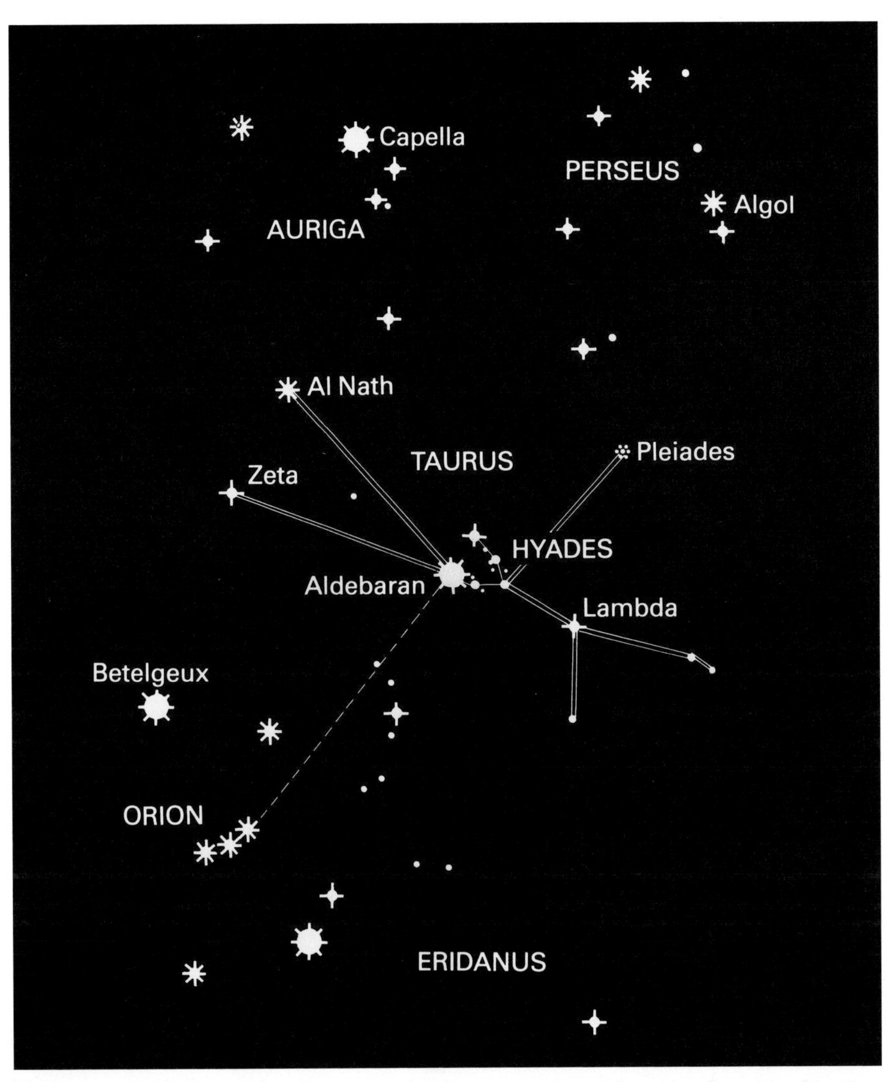

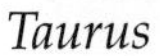
Taurus

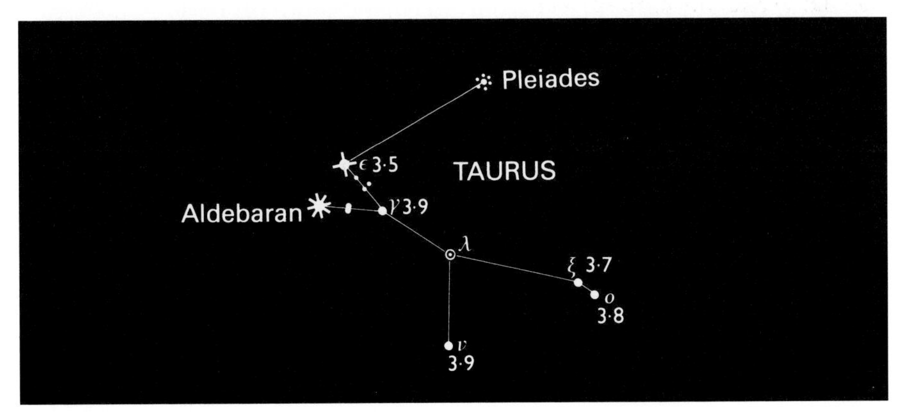

λ *Tauri*

ERIDANUS: *the River*

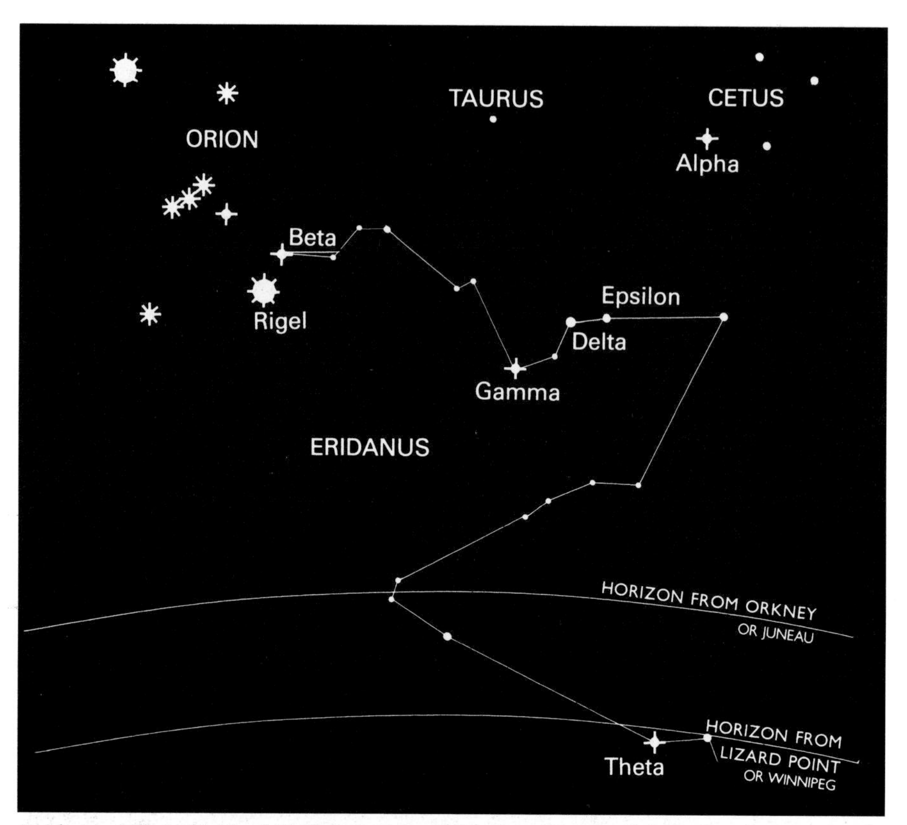

Eridanus. Theta Eridani (2.9) is just invisible from England, but rises in New York. However, the brilliant Achernar (0.5) is much too far south to be seen from anywhere in the United States and so is not shown on the chart.

Eridanus is a long, sprawling constellation which starts near Orion and extends almost to the south celestial pole, so that much of it, including its brightest star (Achernar) is invisible from Britain or the United States. Beta (2.8) is so close to Rigel that it seems to belong more properly to the Orion pattern. Slightly fainter is Gamma (2.9), further west and rather lower down; it is not particularly easy to identify, because there are no obvious pointers to it. Its neighbour Epsilon (3.7), close to the rather brighter Delta (3.5), is however of interest. It is one of the closest stars known, at a mere 10.7 light-years, and is one of the few naked-eye stars which is considerably less powerful than the Sun; it is also smaller, cooler and redder. Yet all in all, it is not too unlike the Sun, and there have been suggestions that it may be a suitable candidate as the centre of a planetary system similar to ours. There is nothing improbable in this, though at the moment we have no proof one way or the other.

The long, straggly line of stars marking Eridanus can be traced down to the southern horizon, and is well placed during January, but on the whole this is a very barren area of the sky.

LEPUS: *the Hare*

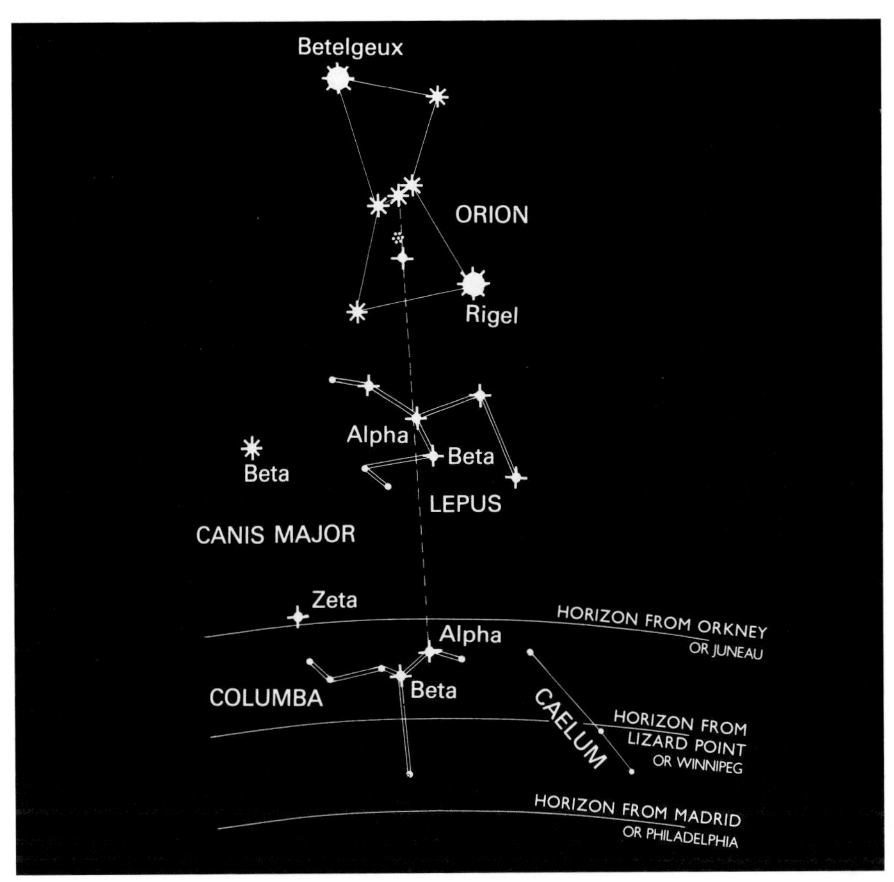

Lepus, Columba and Caelum

Lepus lies below Orion. It has no bright stars, but Alpha (2.6) and Beta (2.8) are easy to locate. The distance between Rigel and Alpha Leporis is about the same as that between Rigel and the Hunter's Belt. There is no well-marked pattern, nor are there any naked-eye objects of note. From Britain, Lepus is always rather low down.

COLUMBA: *the Dove*

Still lower down, below Lepus, may be seen a few stars of another little constellation, Columba (originally Columba Noachi, Noah's Dove). Its leaders, Alpha (2.6) and Beta (3.1), would be easy to find if they were better placed, but are always so near the British or Canadian horizon that they are hard to see; over Scotland they do not rise at all. A line from Alnitak in Orion's Belt, passed through Saiph and continued for some distance, will reach the line of stars which marks Columba.

CAELUM: *the Sculptor's Tools*

There are no notable objects in Columba, and the adjacent group, formerly known as Caela Sculptoris but now shortened to Caelum, is even more obscure. It contains no star as bright as the fourth magnitude, and is always extremely low over Britain and the United States. Caelum was added to the sky by the French astronomer Lacaille in 1752, but barely merits separate identity.

Between them, Orion and Ursa Major serve as direction-finders to most of the constellations visible from northern latitudes, which is an extra reason why winter is an ideal time to start learning the various groups. I did so at the age of seven, when I armed myself with a set of outline star maps and made a pious resolution to identify one new constellation every clear night. I found that it was surprisingly easy, and once a constellation has been located it can usually be found again almost at first glance. Try it, and I think you will see what I mean.

8

Northern stars: the February sky

The two-hour difference between the 10 p.m. charts for January and February results in a considerable shift. Orion has moved well over towards the west, though is still excellently placed, while Capella is now somewhat west of the overhead point.

The maps given here apply to the following times:

1 October:	6 a.m.	(6 hours G.M.T.)
1 November:	4 a.m.	(4 hours G.M.T.)
1 December:	2 a.m.	(2 hours C.M.T.)
1 January:	midnight	(0 hours G.M.T.)
1 February:	10 p.m.	(22 hours G.M.T.)
1 March:	8 p.m.	(20 hours G.M.T.)

They also apply to 6 p.m. on 1 April, though the sky is of course then too light for any stars to be seen.

Taking the northern aspect first, Ursa Major is high up in the north-east, still with its tail pointing in the general direction of the horizon. Follow round the curve of the tail, and you will come to an extremely bright orange star; this is Arcturus in Boötes (the Herdsman), still low down but becoming very conspicuous in the early hours of the morning. Vega is at its lowest. Its position can be checked by using Capella and Polaris – remember that Polaris lies almost exactly half-way between Capella and Vega – and from England or Canada it can usually be found, though from New York it passes below the horizon for a while. Deneb in Cygnus is a little higher, though since it is not nearly so brilliant as Vega it may well be overlooked.

The ever-present Cassiopeia rides high in the north-west, and the other circumpolar groups such as Draco and Cepheus are also on view. So is Perseus, which is not conveniently shown on the hemispherical charts because it spreads from one to the other. The Square of Pegasus is setting, and is unlikely to be seen.

The southern aspect is still ruled by Orion. Eridanus has become low, but Sirius is at its best, and this is also a good time to look for the visible part of Puppis. Aldebaran is starting to descend in the west and will set by 3 a.m., following the Pleiades cluster below the horizon.

To compensate for this, the spring groups are arriving on the scene; in particular there is Leo, the Lion, with the first-magnitude leader, Regulus. A large part of the south-east is occupied by Hydra, the Watersnake, which has

STAR MAP 2

50° N 15 Feb 10 p.m.

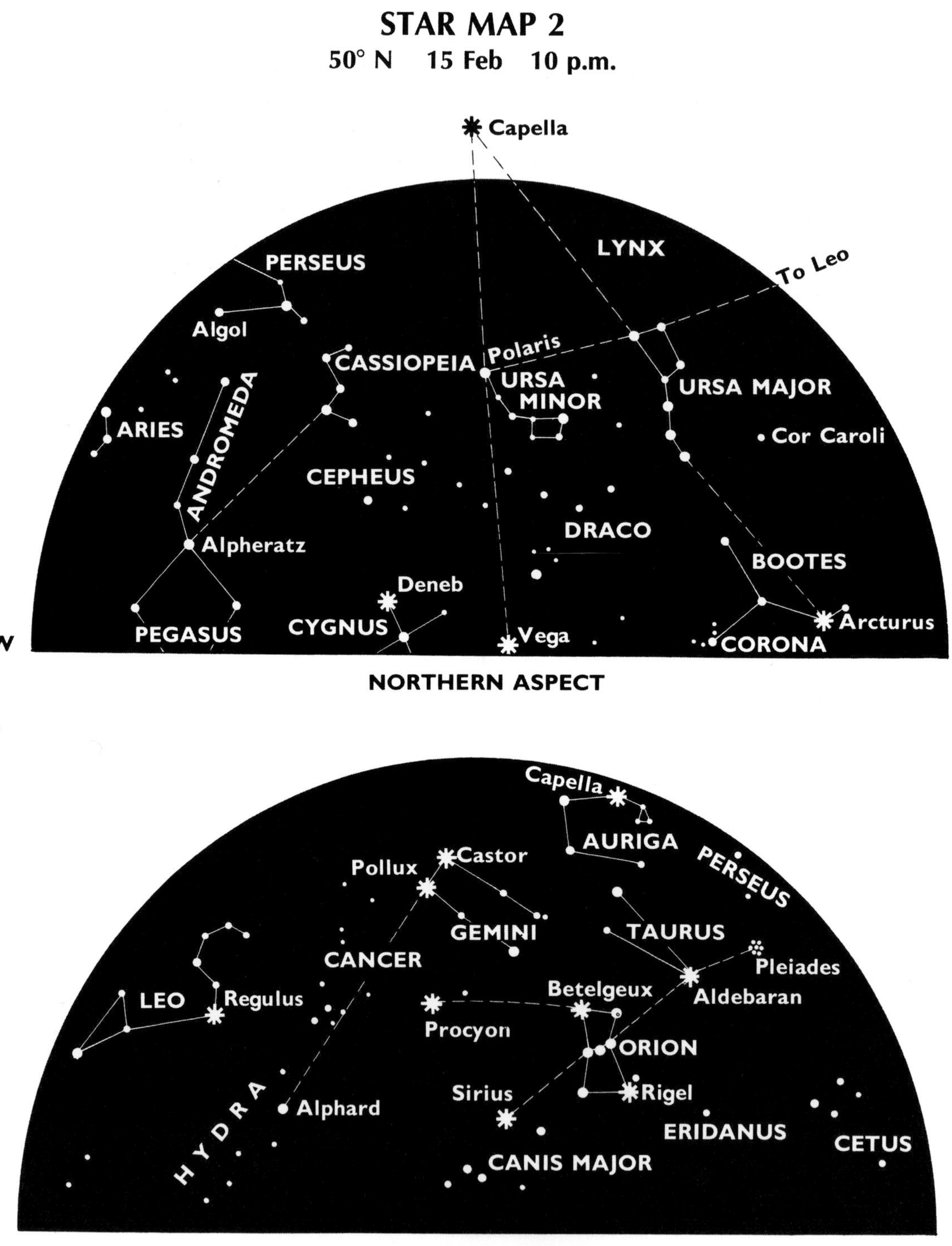

the distinction of being the largest of all the constellations (excluding the now-dismembered Argo) and is also one of the dullest.

One notable feature of the sky during February evenings is the Milky Way, which runs from the northern horizon to the southern, passing through Cassiopeia, Perseus, Auriga, Gemini, Monoceros, and past Sirius into Puppis before being lost to view. City-dwellers are unfortunate here, because they can never hope to see the shining band at all clearly, but from the country, well away from sodium lights and other undesirable features of modern civilization, the effect is magnificent. Of course, moonlight will drown it – but there are always long periods when the Moon is out of the way, to the regret of lunar observers but to the relief of those who are anxious to see the full beauty of the stars.

9

Northern stars: the March sky

Another two-hour shift brings us to the chart for 10 p.m. on 1 March. The maps apply to the following times:

1 November:	6 a.m.	(6 hours G.M.T.)
1 December:	4 a.m.	(4 hours G.M.T.)
1 January:	2 a.m.	(2 hours G.M.T.)
1 February:	midnight	(0 hours G.M.T.)
1 March:	10 p.m.	(22 hours G.M.T.)
1 April:	8 p.m.	(20 hours G.M.T.)

Remember British Summer Time and Daylight Saving Time, when the clocks are put forward one hour during the summer period. Astronomers ignore these, and other time zones such as Mountain Time and Pacific Standard Time.

By 10 p.m. on 1 March Ursa Major is approaching the zenith, though it still lies somewhat east of the overhead point. Arcturus, in line with the curve of the Bear's tail, is already quite high, and its brilliance makes it stand out; it is even brighter than Capella and Vega, though comparisons are difficult during March evenings, because Capella is considerably higher than Arcturus while Vega is still low in the north.

Continuing the curve of the Bear's tail through Arcturus, we come to another first-magnitude star, Spica in Virgo. The relationship is not well shown on the hemispherical charts, but the maps given in Chapter 10 should be helpful. Cassiopeia and Perseus are now in the north-west, while the Square of Pegasus has set.

Orion, together with Aldebaran and the Pleiades, is dropping towards the horizon; Sirius is low, though the Twins and Procyon remain well placed. Leo is prominent in the south; the huge area of Hydra, taking up much of the southern aspect, seems remarkably featureless.

The Milky Way is still conspicuous, though not quite so much in evidence as it was in January or February.

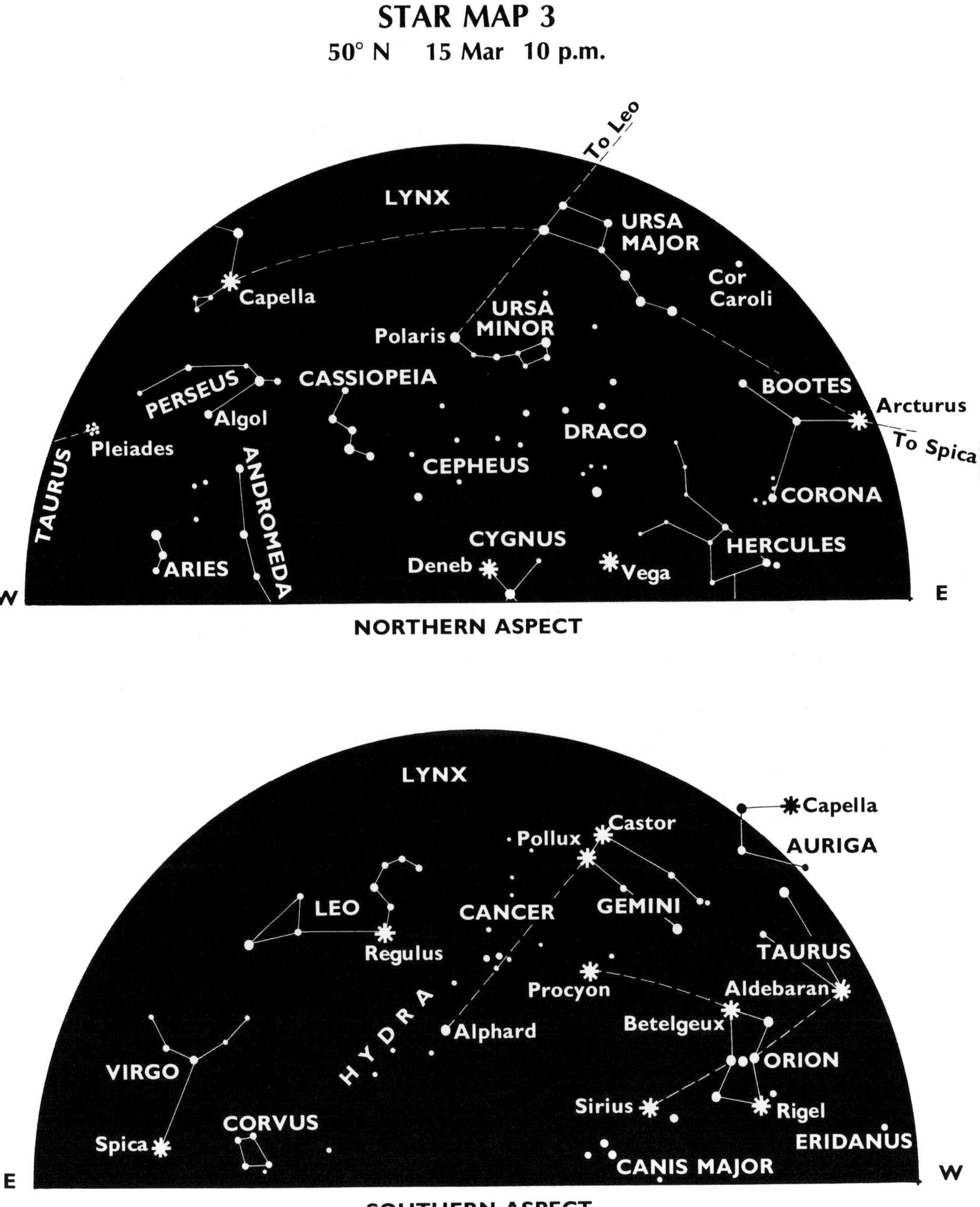
STAR MAP 3
50° N 15 Mar 10 p.m.
To Leo
LYNX
URSA
MAJOR
Cor
Caroli
Capella
URSA
MINOR
Polaris
PERSEUS
CASSIOPEIA
BOOTES
Arcturus
Algol
DRACO
To Spica
Pleiades
TAURUS
ANDROMEDA
CEPHEUS
CORONA
CYGNUS
HERCULES
ARIES
Deneb
Vega
W
E
NORTHERN ASPECT
LYNX
Capella
Castor
Pollux
AURIGA
LEO
CANCER
GEMINI
Regulus
TAURUS
Procyon
Aldebaran
HYDRA
Betelgeux
Alphard
ORION
VIRGO
Sirius
Rigel
CORVUS
ERIDANUS
Spica
CANIS MAJOR
E
W
SOUTHERN ASPECT

10

Northern stars: the April sky

The seasonal change-over in the sky is more or less complete by April evenings. Orion, the symbol of winter, has to all intents and purposes disappeared, though the northernmost part of the constellation still lingers above the horizon and does not actually set until about midnight. Aldebaran and Sirius are barely visible, and of the Hunter's retinue only Procyon and the Twins remain prominent – apart from Capella, high in the west. The spring groups such as Boötes, Leo and Virgo have come well into view.

The April charts hold good for:

1 January:	4 a.m.	(4 hours G.M.T.)
1 February:	2 a.m.	(2 hours G.M.T.)
1 March:	midnight	(0 hours G.M.T.)
1 April:	10 p.m.	(22 hours G.M.T.)
1 May:	8 p.m.	(20 hours G.M.T.)

Ursa Major is practically overhead, which means that Cepheus and Cassiopeia, on the other side of the Pole Star, are reaching their lowest positions in the north, though they never drop low enough to be difficult to find. Vega and Deneb are rising in the north-east, though they have not yet become prominent.

The best way to locate Vega at this time of the year is to use Capella and Polaris as pointers. Unfortunately, Vega and Capella are so far apart in the sky that when using the hemispherical charts, the line joining them appears somewhat 'kinked'. This is due to the way in which the charts have been drawn, but it does not really matter, because both Vega and Capella are so bright. Now that Sirius is out of the reckoning, no star in the evening sky apart from Arcturus is anything like so brilliant – though, as always, beware of planets!

The southern aspect is dominated by Leo, the Lion. Between Leo and the Twins is a rather obscure Zodiacal group, Cancer (the Crab), whose sole claim to distinction is that it contains the naked-eye open cluster Praesepe. From Cancer, the straggly outline of Hydra stretches down to the horizon, passing beneath Leo, while perched on the Watersnake's back is a crow, Corvus, marked by four moderately bright stars which can be quite conspicuous on a clear April evening.

Arcturus and Spica are the other first-magnitude stars on view. At this time of the year Arcturus is visible all through the hours of darkness, while Spica does not set until shortly before sunrise.

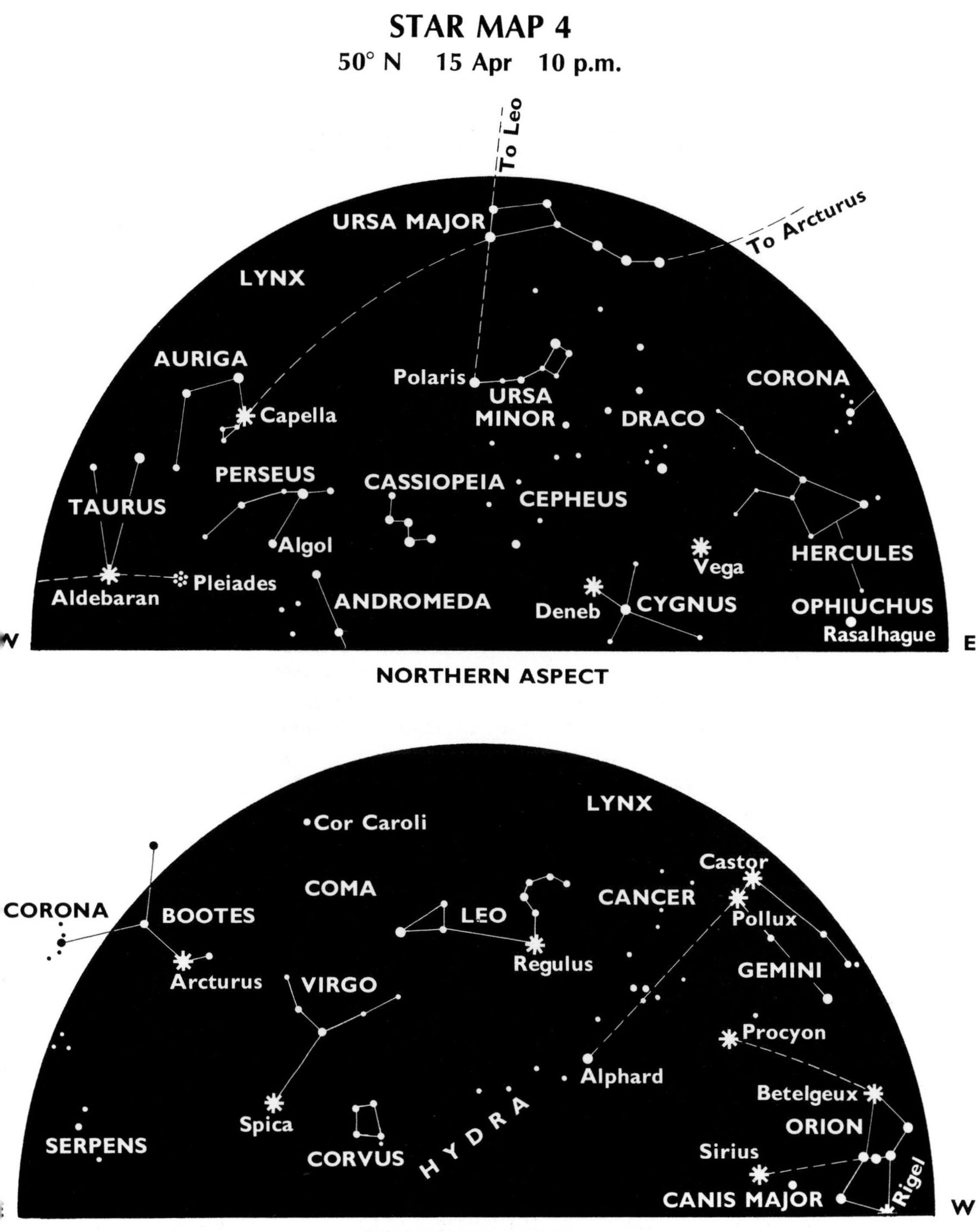
STAR MAP 4
50° N 15 Apr 10 p.m.
To Leo
To Arcturus
URSA MAJOR
LYNX
AURIGA
Capella
Polaris
URSA MINOR
CORONA
DRACO
PERSEUS
TAURUS
CASSIOPEIA
CEPHEUS
Algol
HERCULES
Vega
Aldebaran
Pleiades
ANDROMEDA
Deneb
CYGNUS
OPHIUCHUS
Rasalhague
W
E
NORTHERN ASPECT
LYNX
Cor Caroli
Castor
COMA
CANCER
CORONA
BOOTES
LEO
Pollux
Regulus
Arcturus
VIRGO
GEMINI
Procyon
Alphard
Betelgeux
Spica
ORION
SERPENS
CORVUS
HYDRA
Sirius
CANIS MAJOR
Rigel
W
SOUTHERN ASPECT

BOÖTES: *the Herdsman* and CORONA BOREALIS: *the Northern Crown*

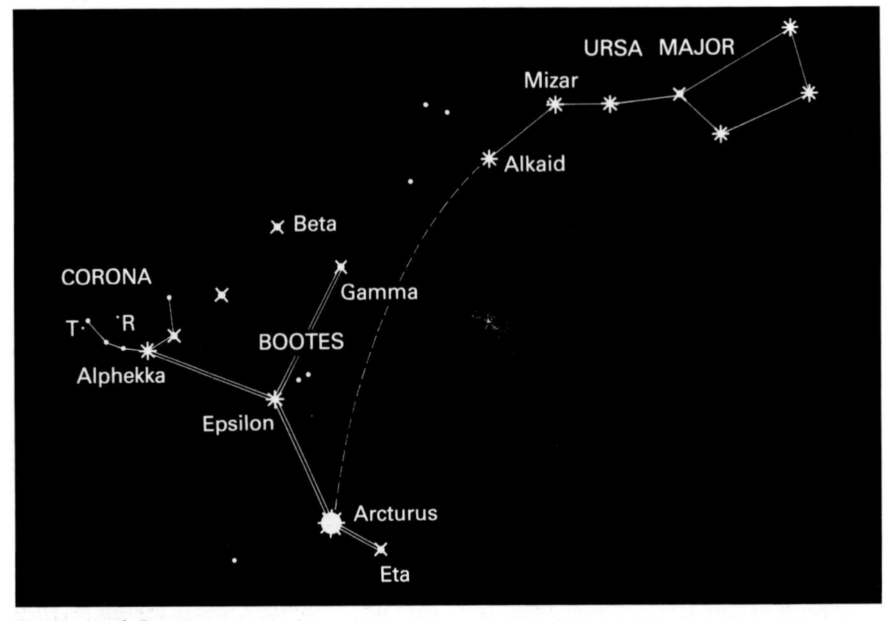

Boötes and Corona

Boötes is, of course, distinguished by the presence of Arcturus, which is of magnitude −0.04. Arcturus, which is a lovely light orange colour, is 36 light-years away, and 115 times as luminous as the Sun. Only three stars (Sirius, Canopus and Alpha Centauri) are brighter, and all these lie to the south of the celestial equator.

The rest of Boötes contains some fairly bright stars, but the pattern seems really to include Alphekka or Alpha Coronae, leader of the neighbouring constellation of Corona Borealis (the Northern Crown). There is a very noticeable 'spoon' outline, made up of Alphekka (2.2) and four of the stars in Boötes: Gamma (3.0), Epsilon (2.4), Arcturus and Eta (2.7). It therefore seems best to include Corona with the map of Boötes.

It is worth taking a closer look at Epsilon, at the top of the spoon-handle. It is of the same colour as Arcturus, but the orange cast is not nearly so evident because the star is so much fainter; to the naked eye it merely looks slightly off-white. It provides a good example of the effect of light-intensity upon observed colour. Delicate hues which can be quite clearly seen in the case of bright objects are totally lost when the intensity drops, which is why the colour of Arcturus is striking to the naked eye, while that of Epsilon Boötes is not.

Corona is one of the few constellations whose outline bears some resemblance to the object after which it is named. There is a small semicircle of stars which could well be interpreted as a crown. Alphekka is almost as bright as Polaris, though the other stars in the pattern are relatively dim, so that mist or moonlight will hide them.

Corona contains the remarkable variable T Coronae (the so-called Blaze Star), which is usually very faint, but which flared briefly up to the second magnitude in 1866 and again in 1946. Also in the bowl of the crown is R Coronae, which is usually on the fringe of naked-eye visibility, but which at irregular intervals falls to a minimum so faint that powerful telescopes are needed to show it. Apparently the cause is nothing more than clouds of soot which accumulate in the star's atmosphere!

VIRGO: *the Virgin*

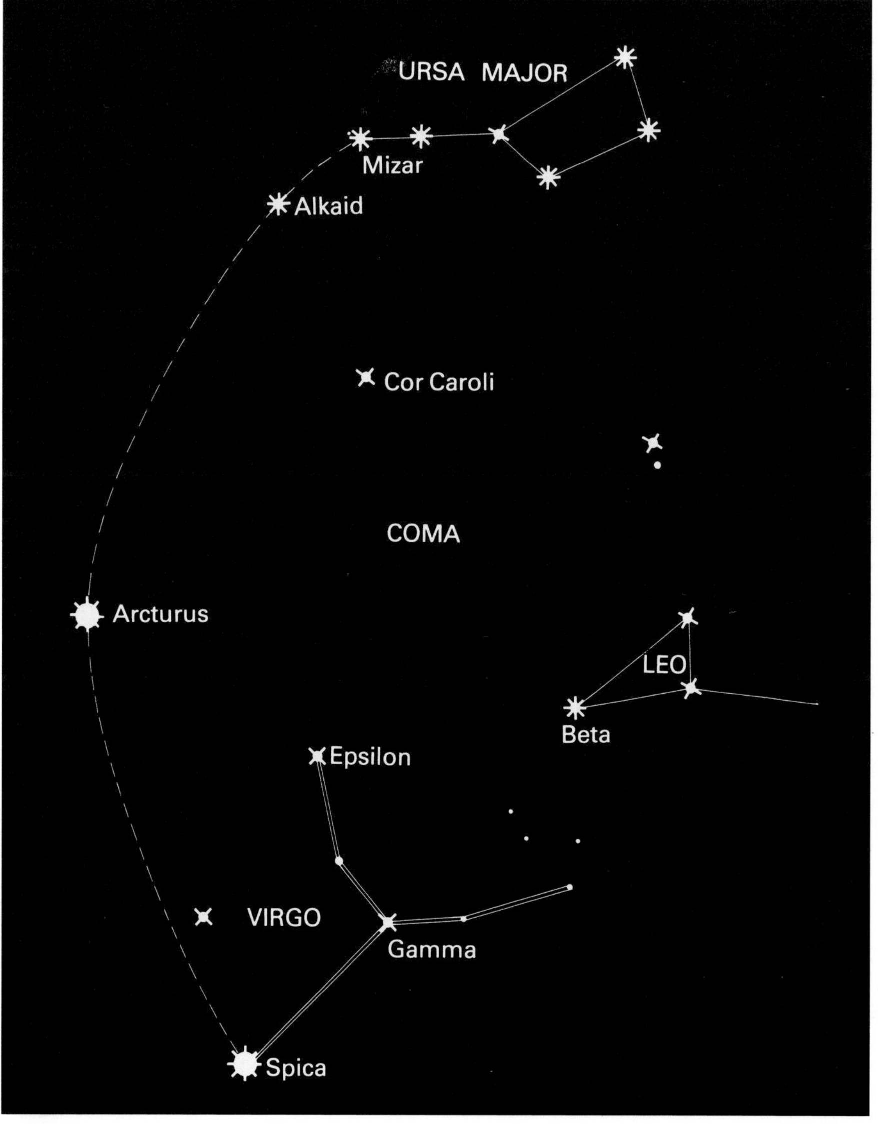

Virgo

A continuation of the line from Ursa Major's tail through Arcturus leads on to Spica, the leader of Virgo. The magnitude of Spica is almost exactly 1, so that in brilliancy it lies midway between Capella and Polaris.

Spica is south of the celestial equator, but rises high over Britain and the United States, so that it is very prominent all through the spring and early summer. The rest of Virgo is made up of a Y-pattern of stars, not unlike the Boötes–Corona spoon, but fainter and more symmetrical. Spica lies at the base of the Y.

Gamma (2.7) is of interest. With the naked eye it looks like a single star, but a small telescope will show that it is a fine binary with two equal components. The only other brightish star in Virgo is Epsilon (2.9), at the top of the Y.

LEO: *the Lion*

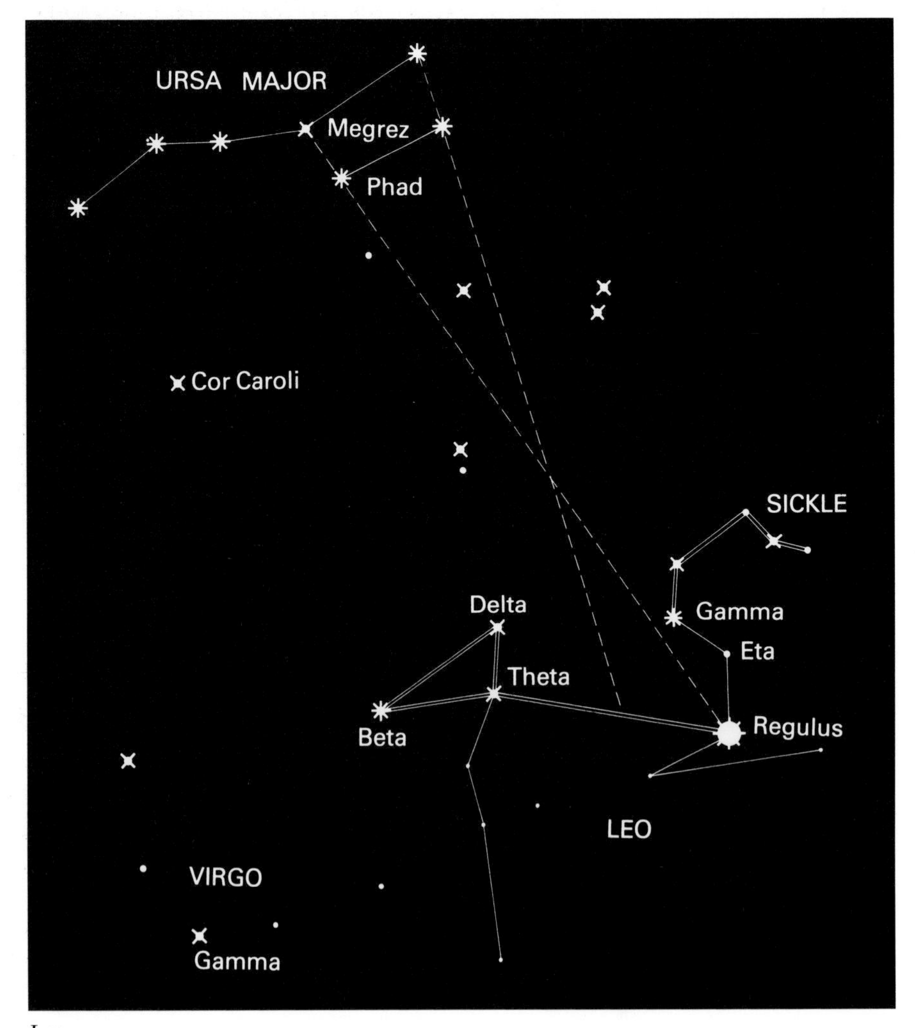

Leo

Leo is certainly the most splendid of the spring groups. It has no star comparable with Arcturus, but it has a very distinctive shape. At midnight in early April it is practically due south, and high up.

As a rough guide, it is sufficient to use the Pointers in Ursa Major, since a direction-line away from the Pole Star will arrive in the middle of Leo. To find Regulus, the brightest of the Lion stars, it is more accurate to use Megrez and Phad, but in any case Regulus stands out because of its brightness. Its magnitude is 1.3, so that it is rather fainter than Spica, but is still very noticeable.

Regulus, known in ancient times as the Royal Star, is white. It lies at the lower end of a conspicuous line of stars arranged in a pattern which is not unlike the mirror-image of a question-mark. This is the famous Sickle of Leo; as well as Regulus it contains the second-magnitude Gamma (2.0), together with several more stars of the third and fourth magnitudes.

The Sickle is the most obvious part of Leo, but there is also a well-marked triangle made up of Beta (2.1), Delta (2.6) and Theta (3.3). Beta or Denebola is interesting, because in ancient times it was said to be of the first magnitude, fully equal to Regulus, though it is now almost a magnitude fainter. It has also been suspected of shorter term variability. Official lists make it about a tenth of a magnitude below Gamma in the Sickle, so that to the naked eye the two should appear virtually equal when they are the same distance above the horizon.

LEO MINOR: *the Little Lion* and SEXTANS: *the Sextant*

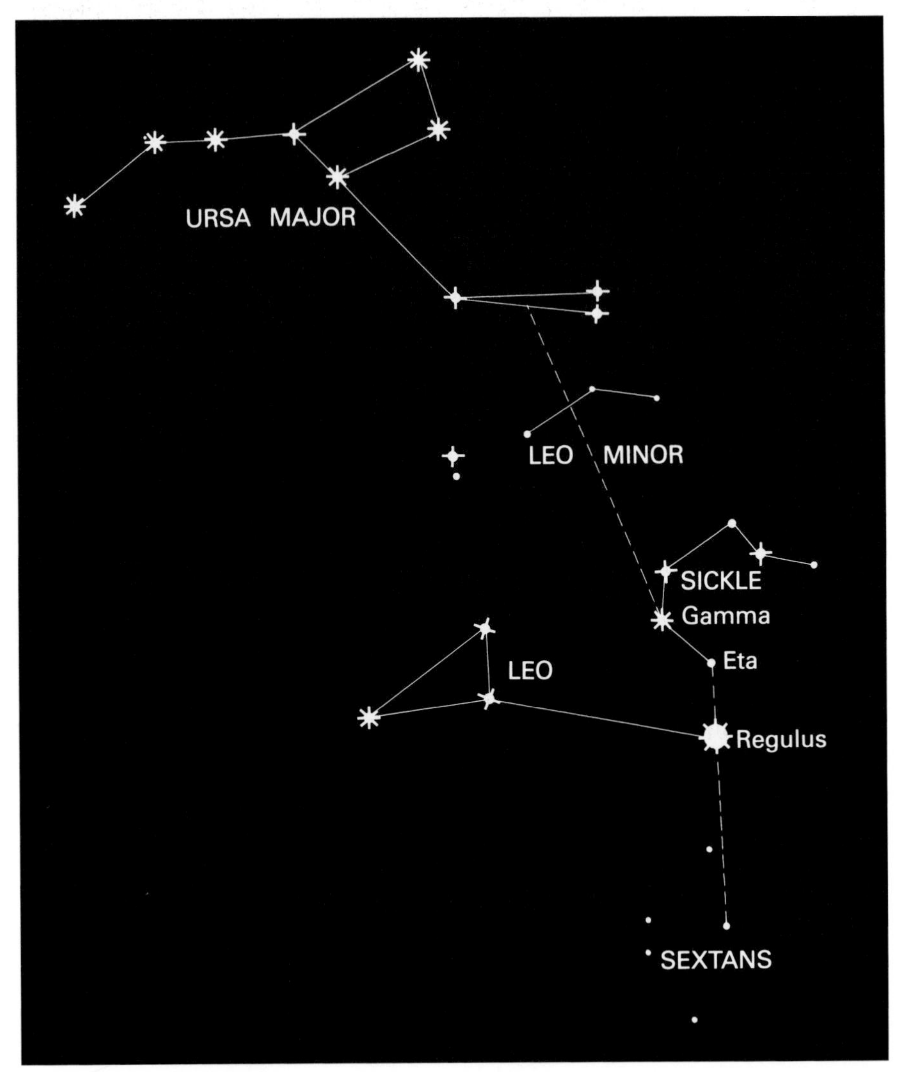

Leo Minor and Sextans

These two constellations near the Sickle of Leo were named by Hevelius in 1690 – though for no good reason, since neither contains any bright stars or obvious outlines. Both may be found by using the Sickle. Leo Minor lies between Gamma Leonis on one side and the third-magnitude triangle of Ursa Major on the other. To locate Sextans, extend a line from Eta Leonis (3.6) through Regulus and continue it for twice the distance between the two. Both groups, particularly Sextans, are so dim and barren that few naked-eye observers will take more than a passing interest in them.

COMA BERENICES: *Berenice's Hair*

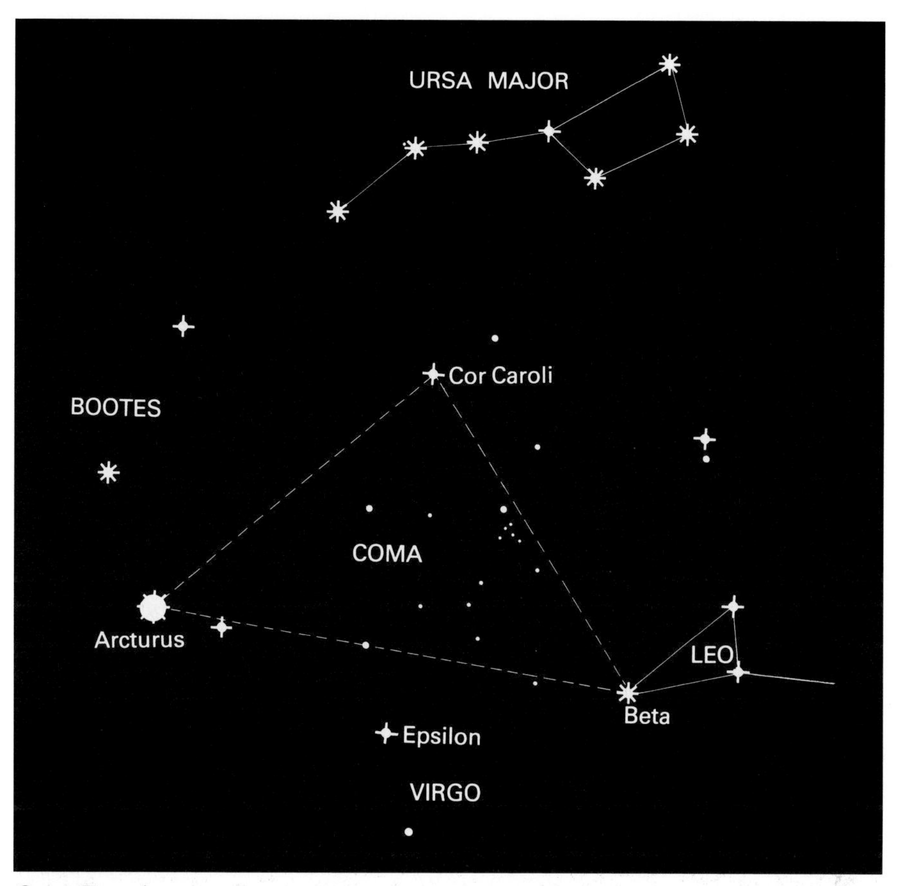

Coma Berenices

Coma is another faint constellation, but a more noteworthy one. It has no star brighter than magnitude 4.5, but there are a great many faint ones, and the whole region gives the impression of being a very large, loose star-cluster.

Coma is easy to find when the sky is clear, but the slightest mist will hide it, so that the entire region will appear blank. The key stars are Arcturus in Boötes, Cor Caroli in Canes Venatici, and Beta Leonis. These three form a large triangle, within which lies most of Coma. The area of numerous dim stars extends south towards Virgo, entering the 'bowl' of the Y. It is, incidentally, very rich in galaxies which are well below naked-eye visibility.

Coma is not an ancient constellation – it was formed by the Danish astronomer Tycho Brahe, who died in 1601 – but the region was studied by the early stargazers, and a legend is attached to it. According to the story, Queen Berenice of Egypt vowed to cut off her beautiful hair and place it in the Temple, provided that her husband returned safely from a dangerous war against the Assyrians. When the king came back unharmed, Berenice kept her promise, so impressing the gods that they transferred the shining tresses to the sky.

CANCER: *the Crab*

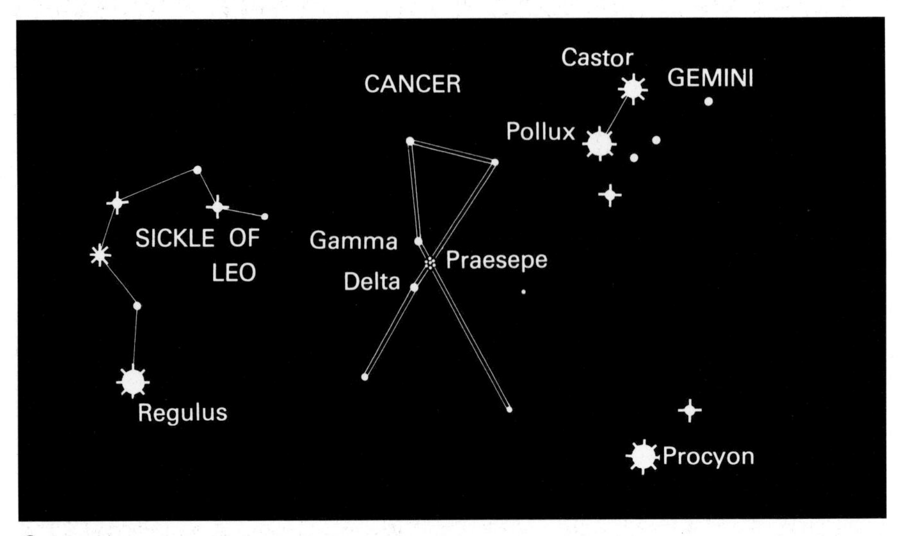

Cancer

The area between Castor and Pollux to the one side and Regulus to the other looks very dull and featureless. It is occupied by Cancer, the celestial crab, which is one of the original constellations, but has no star brighter than Beta (3.5). Most of it is contained within the triangle formed by Regulus, Pollux and Procyon.

Identification is not difficult, because although Cancer is faint it has at least a definite shape – it looks rather like a very dim and ghostly Orion. Moreover, there is one object of real interest. This is Praesepe, an open star-cluster inferior only to the Pleiades.

Praesepe, Messier No. 44, is clearly visible with the naked eye against a dark sky. Moonlight will conceal it, and so will any trace of mist, but under good conditions it is plain enough, and during April evenings it is conveniently high up. Look at the point half-way between Regulus and Pollux, and you will make out two faint stars of Cancer, Delta (4.2) and Gamma (4.7). Between the two, slightly west of a line joining them, will be seen the faint shimmer of Praesepe, which has been aptly nicknamed the 'Beehive'. No individual stars can be made out except with optical aid, but all the same Praesepe is a fine example of a galactic cluster. If you can see it, you may be sure that the air is clear and transparent.

There is nothing else of naked-eye interest in Cancer, but it lies in the Zodiac, so that bright planets pass through it from time to time.

HYDRA: *the Watersnake*

It has been said, graphically if inelegantly, that Hydra is 'a lot of nothing'. It sprawls across the southern aspect during April evenings, beginning not far

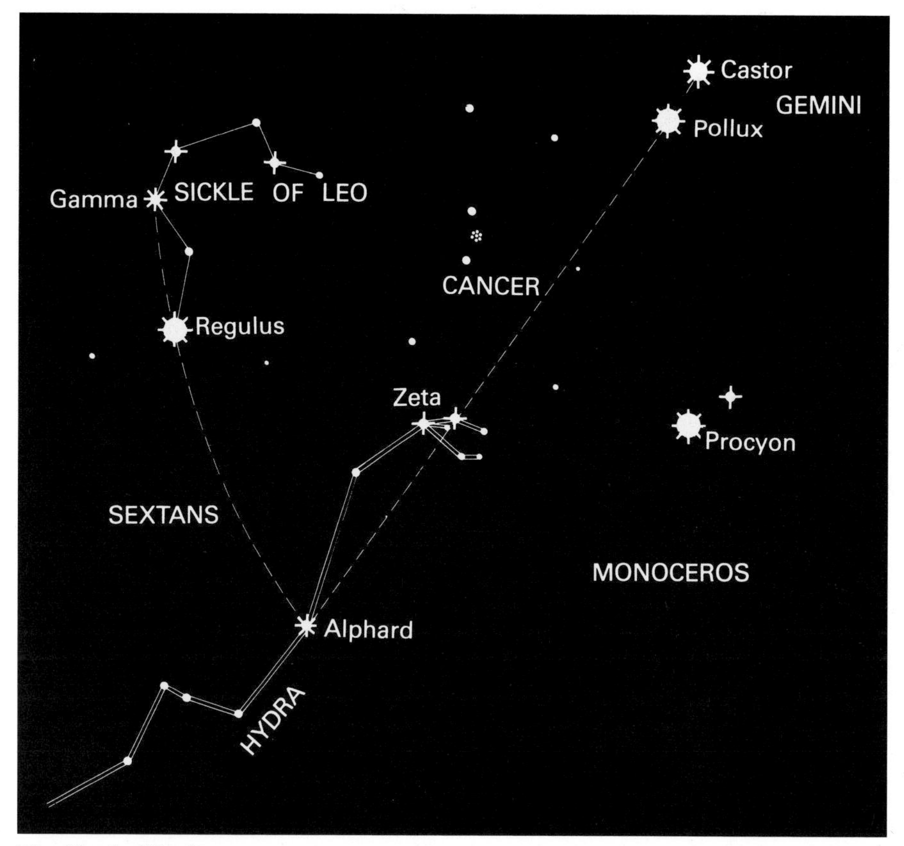

The Head of Hydra

from Procyon and ending below Spica. It contains only one bright star, Alphard or Alpha Hydrae (2.0), and there is a remarkable dearth of interesting objects.

Alphard has been called 'the Solitary One'. There are no conspicuous stars anywhere near it, and for this reason it is easy to find, particularly since it rises to a fair altitude over Britain and North America. It is a Red Giant, 85 light-years away and well over a hundred times as powerful as the Sun. Its colour is detectable with the naked eye, while binoculars bring it out strongly. Alphard has been suspected of variability, but comparison stars are lacking and it is therefore difficult to estimate.

Castor and Pollux act as pointers to Alphard; the line crosses part of Cancer on the way, and also passes through the Watersnake's head, which is made up of four stars fairly close together (Zeta, magnitude 3.1, is the brightest of them). Gamma Leonis and Regulus, in the Sickle, can also be used to locate Alphard if the Twins are out of view.

The rest of Hydra is made up of a line of stars below Leo and Virgo; one, Gamma, is of the third magnitude. Close to Gamma is the Mira variable R Hydrae, which has a range of 4 to 10 and a period of 386 days; it is well above the naked-eye limit when near maximum, but its low altitude makes it difficult to locate from northern countries.

CORVUS: *the Crow*

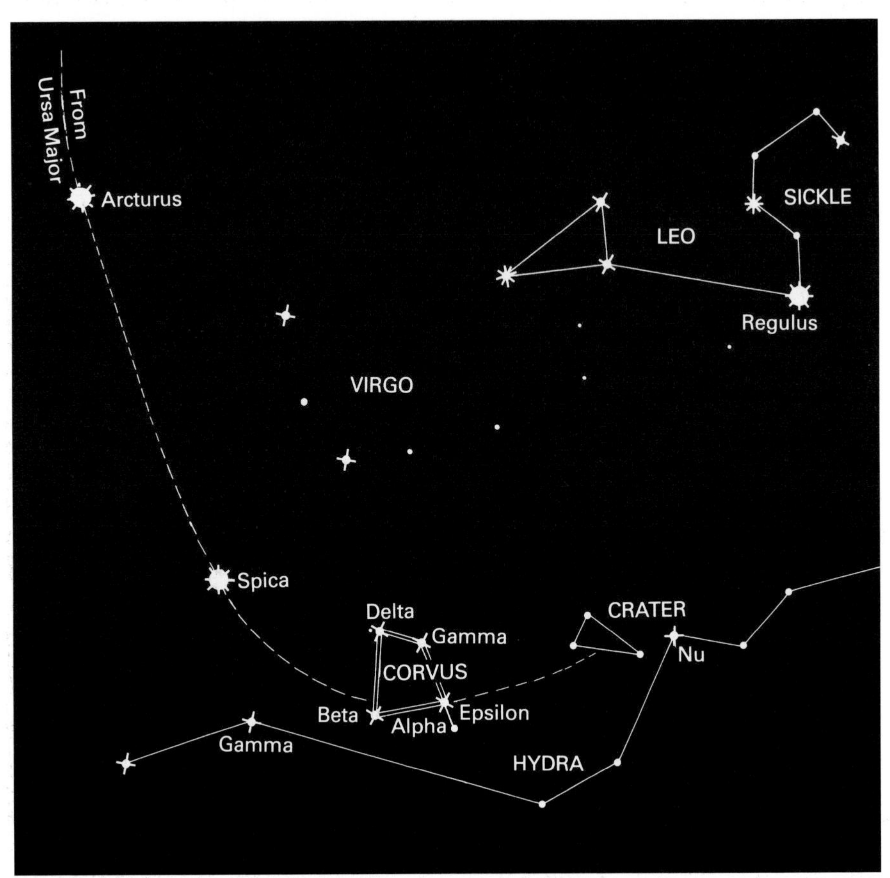

Corvus, Crater and the Tail of Hydra

Corvus lies well south of the equator, but when at its highest, as during late evenings in April, it is prominent enough. The best way to locate it is to continue the curve from Ursa Major's tail through Arcturus and Spica. The curve has to be twisted somewhat, but Corvus is recognizable because it lies in so barren an area. The quadrilateral is formed by Gamma and Beta (each 2.6), Delta (2.9) and Epsilon (3.0). Alpha Corvi, close to Epsilon, is only of magnitude 4.2 even though Bayer gave it the first letter of the Greek alphabet. There are no notable objects in the group.

CRATER: *the Cup*

The second of the small constellations adjoining Hydra is Crater, one of the original groups even though it has no star brighter than magnitude 3.5. It is not easy to identify, because there are no convenient pointers to it. It occupies the space between the quadrilateral of Corvus and the third-magnitude star Nu Hydrae, and consists of a small, dim triangle. The best way to find it is probably to curve up a line from the two lower stars of Corvus, as shown in the diagram opposite. Crater is entirely unremarkable.

CENTAURUS: *the Centaur*

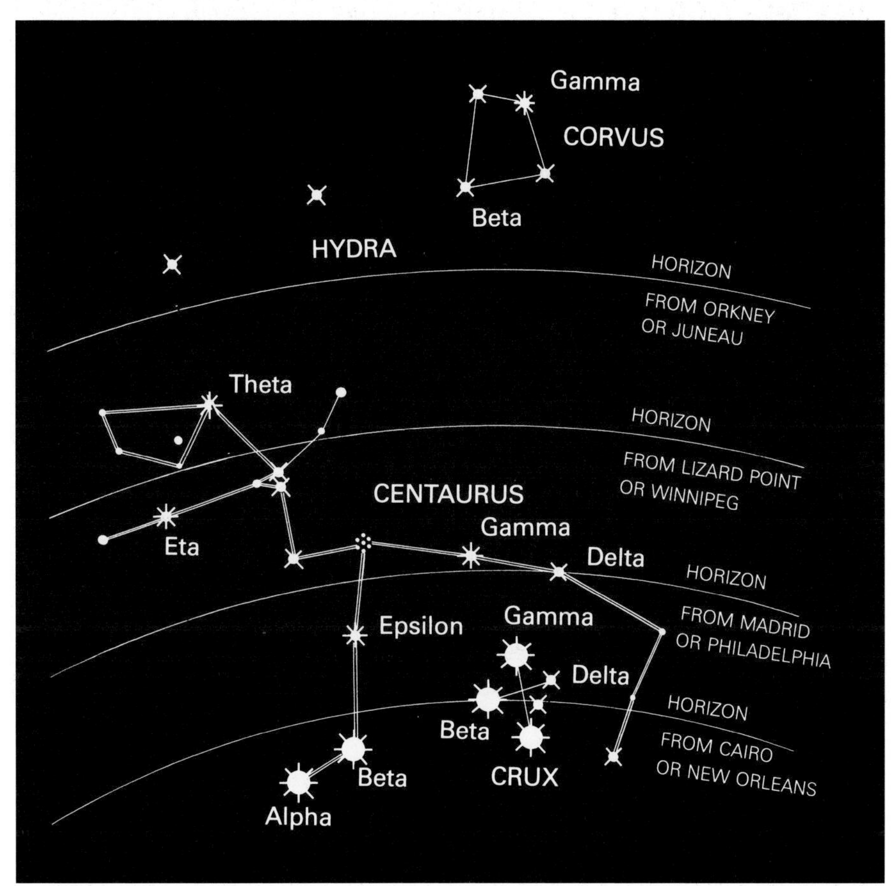

Centaurus

Centaurus is a grand constellation, but almost all of it lies too far south to be seen from Britain or the United States. One of its main stars, Theta (1.9), grazes the horizon from southern England, and various observers claim to have seen it, though I have never done so myself. I have, however, seen it excellently from Arizona, together with other stars in the northern part of the constellation, and it should be seen from the New York area when at its highest. It is situated almost in a line with two of the stars in Corvus, Gamma and Beta. People who live on the south coast of Devon or Cornwall may care to look for it; midnight in early April is the best possible time, though the chances of success are slight.

11

Northern stars: the May sky

The May chart is the first in which Orion is absent altogether. The brilliant Hunter has vanished below the horizon, and of his retinue only Capella and the Twins remain. These, fortunately, can be located by using Ursa Major, so that from a direction-finding point of view the loss of Orion is not disastrous.

The times for which the new chart holds good are:

1 January:	6 a.m.	(6 hours G.M.T.)
1 February:	4 a.m.	(4 hours G.M.T.)
1 March:	2 a.m.	(2 hours G.M.T.)
1 April:	midnight	(0 hours G.M.T.)
1 May:	10 p.m.	(22 hours G.M.T.)

Of course, it is also valid for 1 June at 20 hours G.M.T., but the sky is then too light for stars to be seen. Ursa Major is more or less overhead, which means that Cassiopeia is at its lowest, though still well above the northern horizon. Arcturus is high up and Spica almost due south, with Regulus and the Sickle very prominent in the south-west. Castor and Pollux are dropping; Cancer, with the 'Beehive' cluster Praesepe, can be seen easily enough, together with the whole of Hydra. The Corvus quadrilateral is quite conspicuous in the south when there are no horizon mists.

Vega has reached a respectable altitude in the north-east, and Deneb in Cygnus is becoming noticeable, though Altair in Aquila, the third of the brilliant summer stars, has yet to appear. Much of the south-eastern aspect is taken up by the large, dim groups of Hercules, Ophiuchus (the Serpent-Bearer) and Serpens (the Serpent), which are at their best during the summer. Another dim constellation is Libra, the Balance, which is notable only because it lies in the Zodiac.

Going back for a moment to our circumpolar groups, it is worth noting that Draco is now well placed, and the Dragon's head, close to Vega, is easy to find even though its stars are not bright.

The Milky Way is not so well placed as on earlier charts. It is lower in the sky, and of course the period of darkness is shorter. However, it is still magnificent whenever the Moon is absent. Lastly, do not forget Scorpius, the Scorpion, which is just rising. Its leading star, the fiery red Antares, comes into view around midnight and is very conspicuous during the early hours of the morning. It is not on the May charts, but appears on those which follow.

STAR MAP 5

50° N   15 May   10 p.m.

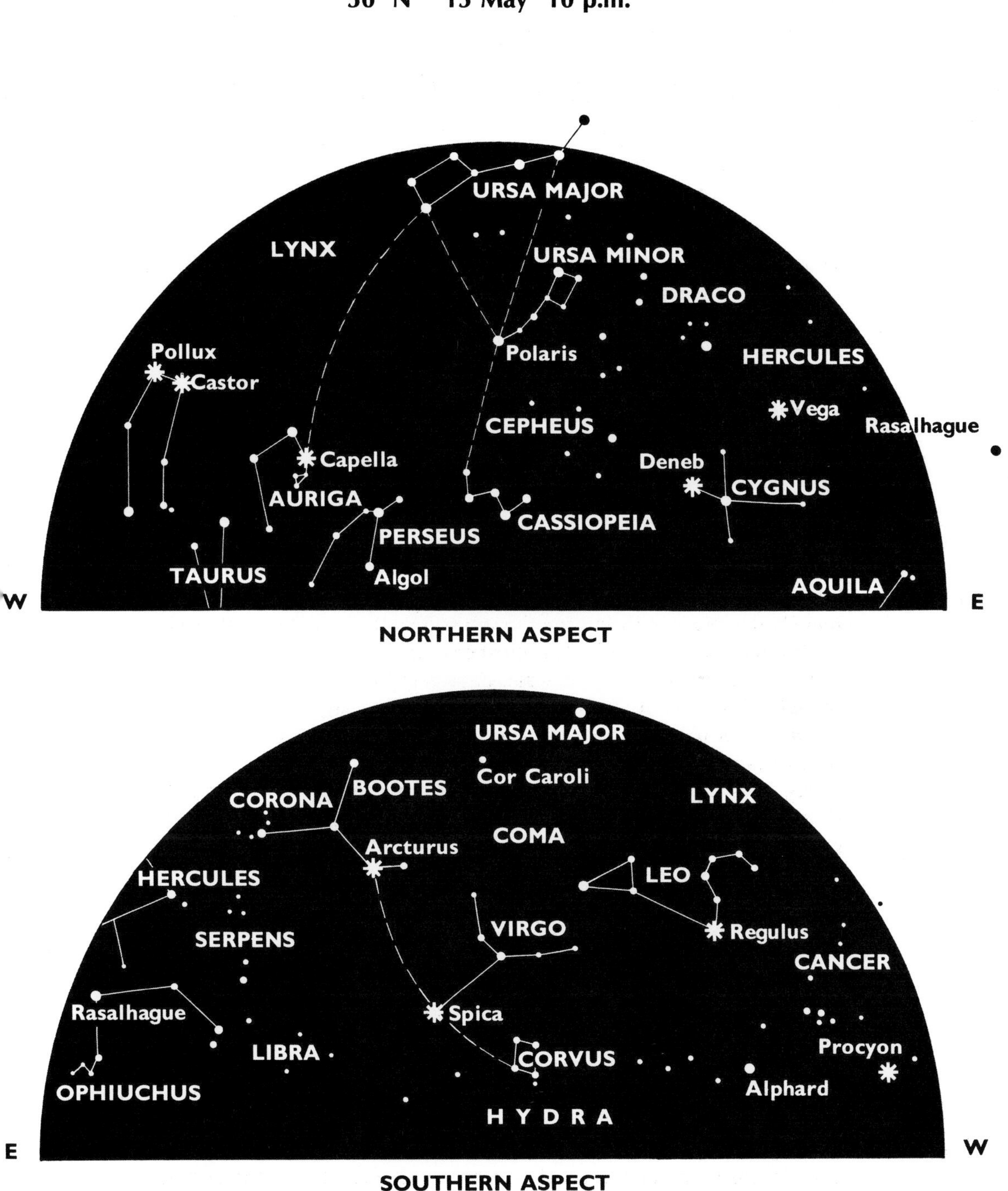

12

Northern stars: the June sky

June can be a somewhat depressing month for the would-be stargazer who is disinclined to keep late hours. Official twilight lasts all night over Britain, and when the Moon is near full there is no proper darkness at all.

Timings for the June chart are as follows:

1 February:	6 a.m.	(6 hours G.M.T.)
1 March:	4 a.m.	(4 hours G.M.T.)
1 April:	2 a.m.	(2 hours G.M.T.)
1 May:	midnight	(0 hours G.M.T.)
1 June:	10 p.m.	(22 hours G.M.T.)
1 July:	8 p.m.	(20 hours G.M.T.)

Ursa Major has shifted to a point rather westward of the zenith, but is still very high up; Cassiopeia has gained altitude in the east, while all three first-magnitude stars of the 'Summer Triangle' (Vega, Deneb and Altair) are also easterly. Arcturus is almost due south, very high up, while Leo and Virgo are descending westward; Corvus is still quite prominent when the sky is clear and dark. In the west, Castor and Pollux are very low, and Capella is not much in evidence. The large, faint groups of Hercules, Ophiuchus and Serpens occupy much of the south-east.

However, there is one brilliant red newcomer: Antares in Scorpius. Deneb and Vega act as rough pointers to it, but this is not well shown on the hemispherical charts, because Deneb and Vega have to be drawn on the northern map and Antares on the southern.

The Milky Way is not seen to advantage in June, partly because it is relatively low down but mainly because the sky is not dark enough.

STAR MAP 6

50° N 15 Jun 10 p.m.

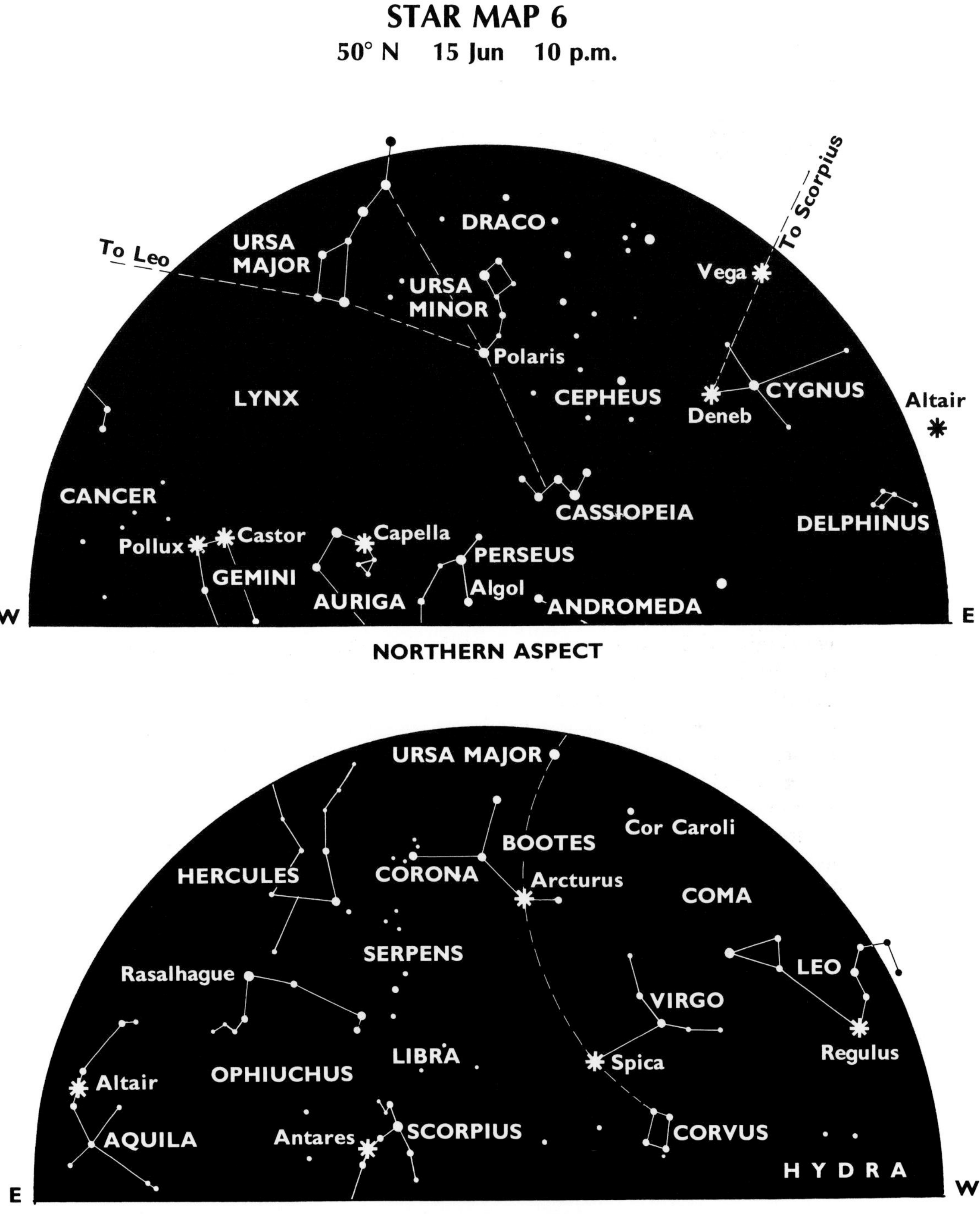

13

Northern stars: the July sky

The chart for July is a logical development of that for June. The spring groups such as Leo and Virgo are setting, while the autumn constellation Pegasus has made its appearance in the east. It is rather unfortunate that large, faint groups occupy so much of the southern aspect, but at least Antares is there, shining redly from the low south, while the richest part of the Milky Way is also on view.

The charts apply to:

1 March:	6 a.m.	(6 hours G.M.T.)
1 April:	4 a.m.	(4 hours G.M.T.)
1 May:	2 a.m.	(2 hours G.M.T.)
1 June:	midnight	(0 hours G.M.T.)
1 July:	10 p.m.	(22 hours G.M.T.)
1 August:	8 p.m.	(20 hours G.M.T.)

Ursa Major is very high up but now lies west of the zenith, which means that Cassiopeia has gained altitude in the north-east. Arcturus is still much in evidence, but Spica has become very low.

A glance at the two hemispherical maps brings home another of their limitations. As we know, the Great Bear's tail points towards Arcturus, and this is obvious when you look at the sky. The hemispherical maps confuse it badly, because the Bear and Arcturus have to be shown on different charts. Unfortunately there is no solution without reverting to the usual circular maps, which are certainly more accurate but are not really of much use in helping the beginner to find his way around the sky.

Vega continues to shine down almost from the overhead point, but Capella, in the north, is so low that it is difficult to see; it should be glimpsed from Britain and Canada, but not from the New York area, where it will have dipped below the horizon. Regulus has more or less vanished in the west, while the Square of Pegasus, Andromeda and Perseus are just about observable in the east. I do not propose to say more about Pegasus here, since it so definitely belongs to the autumn, but by midnight in July it is very much in evidence.

Vega, Deneb and Altair make up a large triangle, though they are of course in different constellations. In a *Sky at Night* television programme long ago (it must have been around 1958) I referred to this as the 'Summer Triangle', a term which everyone now seems to use, though it applies only to the northern hemisphere of the Earth – Britain's summer is Australia's winter.

STAR MAP 7

50° N 15 Jul 10 p.m.

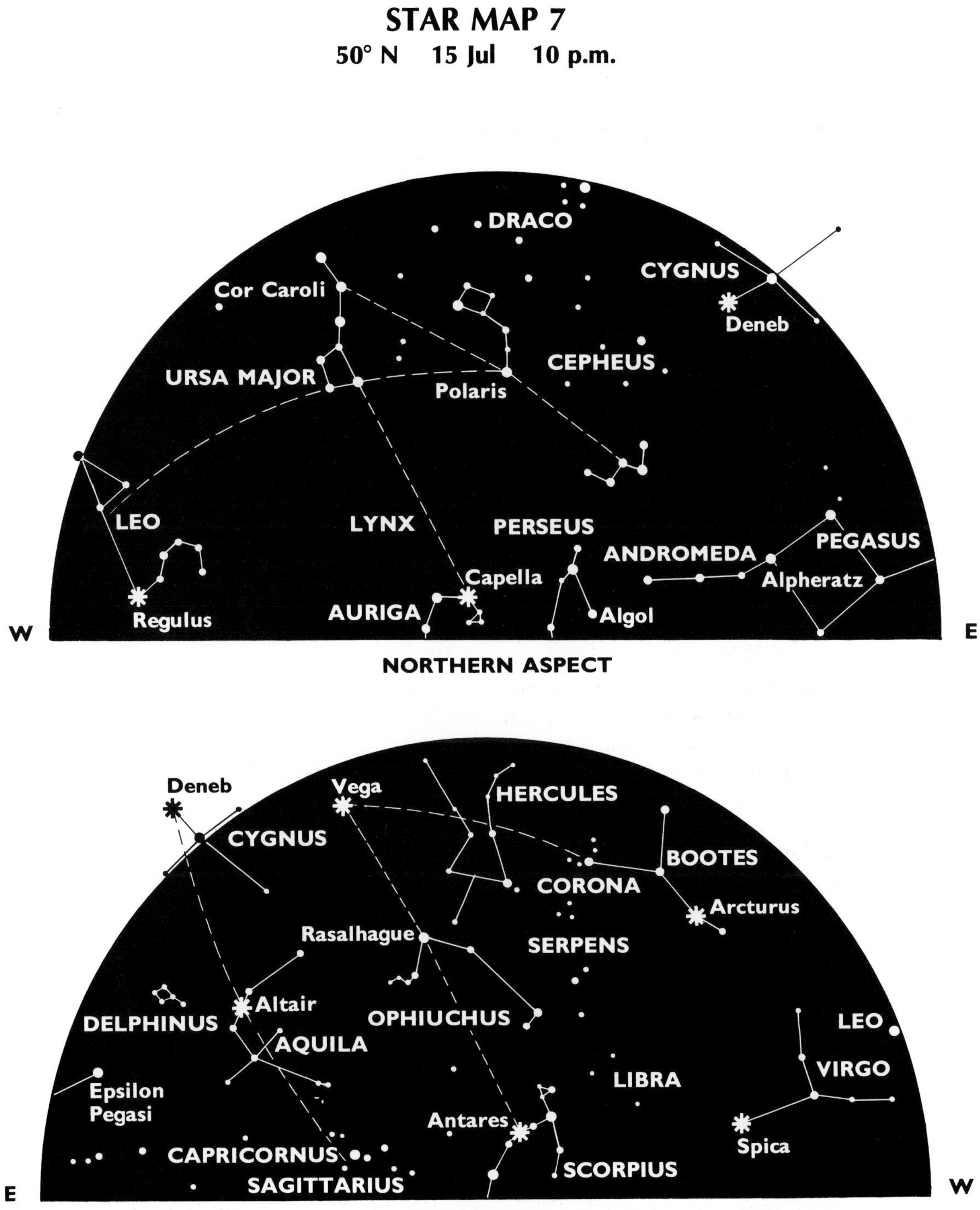

LYRA: *the Lyre*

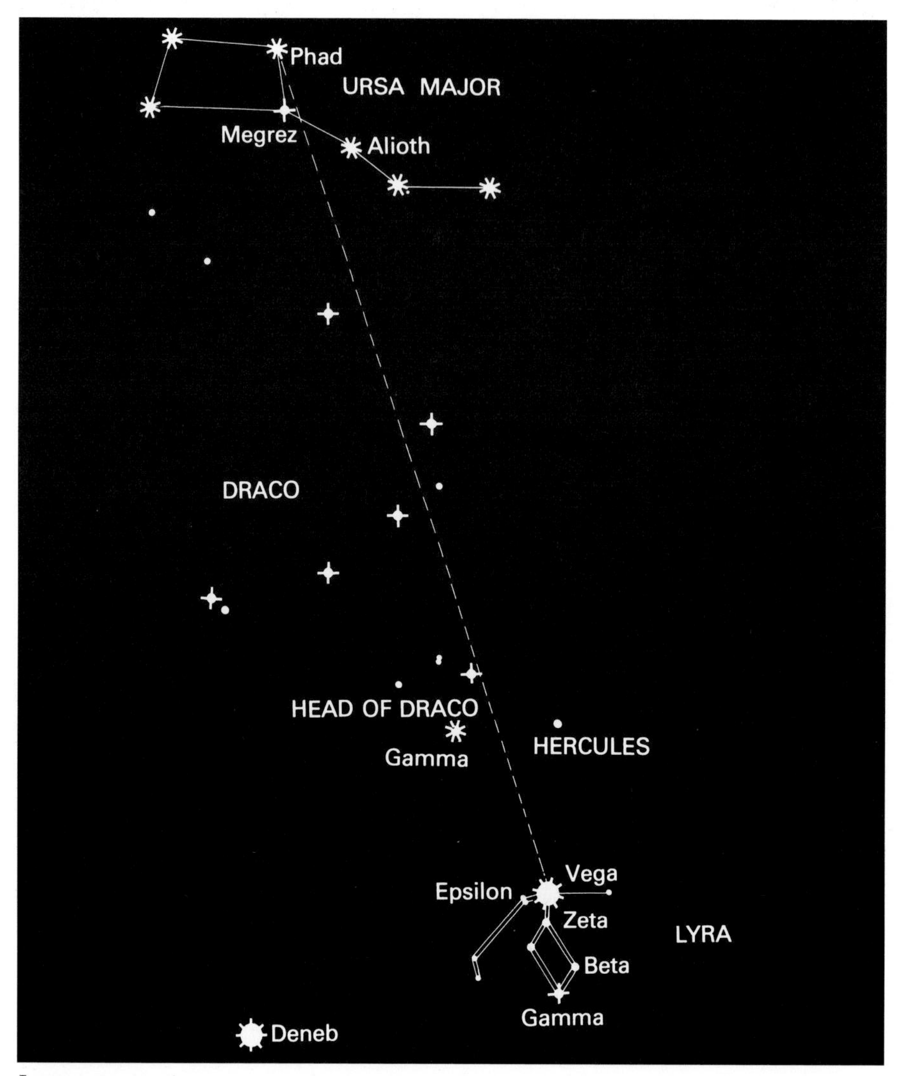

Lyra

Vega is quite unmistakable. It is the equal of Capella, and of all the stars visible from Britain and North America only Sirius and Arcturus outrank it. It lies at the zenith during summer evenings, and its brilliance, together with its steely-blue colour, makes it stand out at once.

This is fortunate, because there are no convenient pointers to it. It may be found from Ursa Major, by the rather awkward method of starting at Phad and directing a guide-line between Megrez and Alioth; it also lies in a line with Capella and Polaris, though this is not much help when Capella is virtually out of view. At any rate, I doubt whether anyone will have much trouble in finding it. If you look upward on a summer evening and see a very bright star, it can only be Vega.

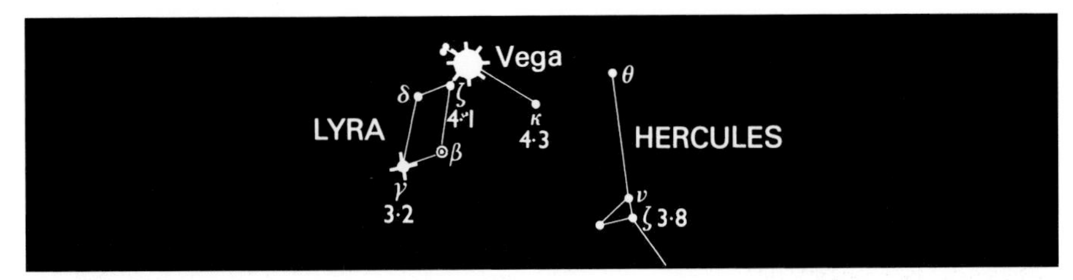

β Lyrae

Its blue colour shows that it is hotter than the Sun. It is also much more luminous, and is the equal of 50 Suns; the distance is 26 light-years, so that on the cosmical scale it is a fairly near neighbour of ours.

Interesting observations of it were made in 1983, from equipment carried aboard IRAS, the Infra-Red Astronomical Satellite, which was launched in January and continued to send back data until the late autumn. Vega was found to be accompanied by material which was not hot enough to shine, but was emitting at long wavelengths. There have been suggestions that this indicates the presence of planet-forming material, or even a fully fledged planetary system. It would be unwise to jump to any conclusions, but the possibilities are there. However, Vega is using up its energy more quickly than the Sun and is not nearly so old, so that even if planets exist it is hardly likely that any Earth-type life has developed there.

The rest of Lyra is not conspicuous, but it contains more than an average share of important objects. First there is the famous naked-eye double Epsilon Lyrae, which lies close to Vega. Normal-sighted people will be able to see that it is made up of two 4½ magnitude stars side by side. With a telescope, each component is again double, so that Epsilon Lyrae is a quadruple system, though the naked-eye observer will have to be satisfied with a straightforward division into two.

The twins of Epsilon Lyrae are genuinely associated, but the two bright pairs are at least a million million miles apart. Clear skies are needed for both to be seen, but Vega provides a convenient guide.

The rest of Lyra is made up mainly of a quadrilateral of stars, of which the brightest is Gamma (3.2). Beta Lyrae is the celebrated eclipsing binary. Its light-curve shows that changes in brightness are always going on. Starting from maximum (3.4) it fades down almost to the fourth magnitude; it recovers to its maximum of 3.4, and then fades again, this time down to 4.3. There are alternate deep and shallow minima, so that the behaviour is unlike that of Algol. Gamma makes a good comparison star, together with Zeta (4.1) and Kappa (4.3).

R Lyrae, in line with Vega and Epsilon, is a genuine variable. It is semi-regular, with a rough period of about 48 days and a range of magnitude from 4 to 5. The comparison stars are Eta and Theta Lyrae (each 4.5) and 16 Lyrae (5.2). Like most semi-regulars, R Lyrae is a red supergiant, but it is too faint for its colour to be noticed with the naked eye.

The best known of all planetary nebulae, M. 57, lies directly between Beta and Gamma, but since it is not visible without a telescope it need concern us no more here.

CYGNUS: *the Swan*

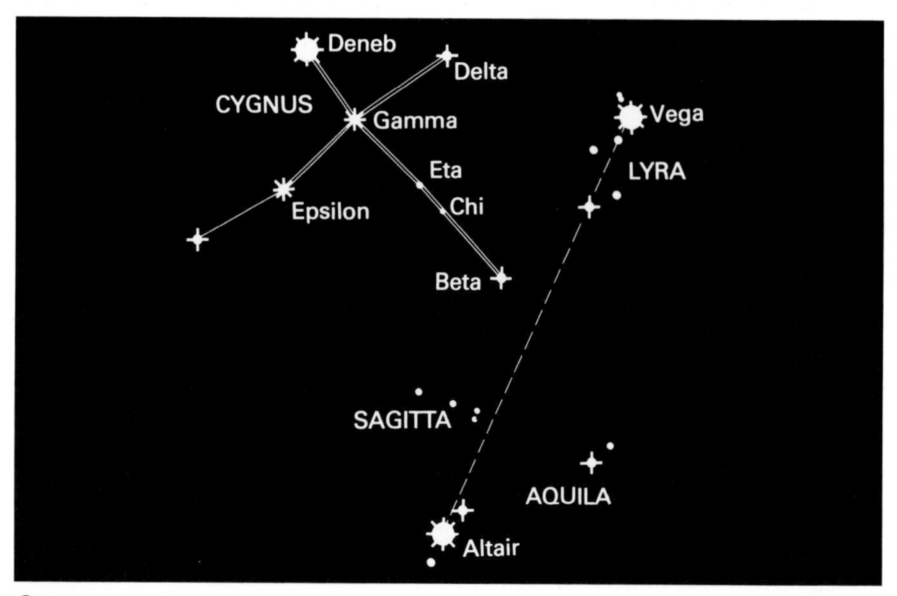

Cygnus

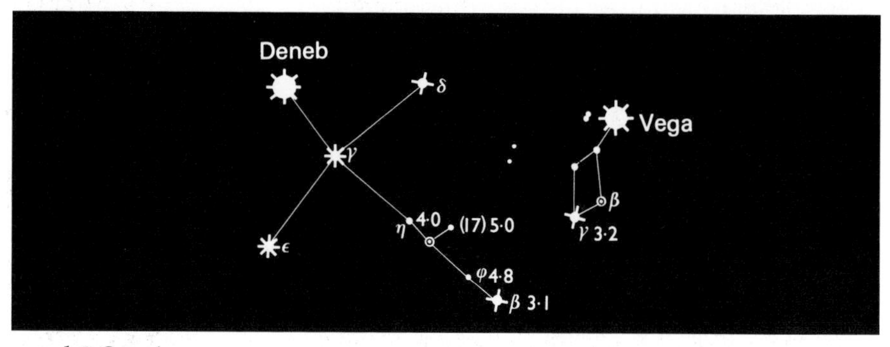

χ and β Cygni

The celestial swan is often, and more appropriately, termed the Northern Cross, since its X-pattern is striking. Deneb, the leading star, is of magnitude 1.2. This is over a magnitude fainter than Vega, but appearances are deceptive. Deneb is one of our 'searchlights', perhaps 70,000 times as luminous as the Sun, and so remote that we are now seeing it as it used to be when the Romans still occupied England.

There are no obvious pointers to Deneb, but it lies east of Vega and should be easy to locate. The other stars of the X are Gamma (2.2), Delta (2.9), Epsilon (2.5) and Beta (3.1). Of these Beta, often still known by its proper name of Albireo, is the most noteworthy. It is fainter than the rest, and further from the middle of the cross so that it spoils the symmetry of the pattern, but to compensate for this it is a glorious telescopic double, with a yellow primary and a vivid blue companion. A good way to identify it is to remember that it lies not far off the line joining Vega and Altair.

Half-way between Albireo and Gamma lies Eta Cygni (3.9). At times a second star may be seen close to it. This is Chi Cygni, a Mira-type variable with a period of 407 days and an exceptional range in brightness. It has been known to reach magnitude 3.6, brighter than Eta, though at other maxima it remains below 4; near minimum it becomes so dim (magnitude 14) that large telescopes are needed to show it at all. It is visible with the naked eye for only a few weeks during its cycle, so that normally the naked-eye observer will not see it, but it is always worth looking for. Yearly almanacs give the times when it should be on view. It is very red, though once again the low light-intensity means that its colour cannot be seen without optical aid – though it is worth noting that at infra-red wavelengths it is one of the brightest objects in the entire sky.

The Milky Way passes straight through Cygnus, running from Cassiopeia in the north-east and continuing into Aquila. The shining band is divided into two some distance south of Gamma, because of the presence of dark nebulae; there are many glorious star-fields, and altogether Cygnus is one of the richest regions in the northern sky.

AQUILA: *the Eagle*

Altair, leader of the Eagle, is the third member of the unofficial Summer Triangle. Unlike Deneb or Vega, it is not circumpolar from anywhere in Britain or North America, but in July it is high up, somewhat east of south.

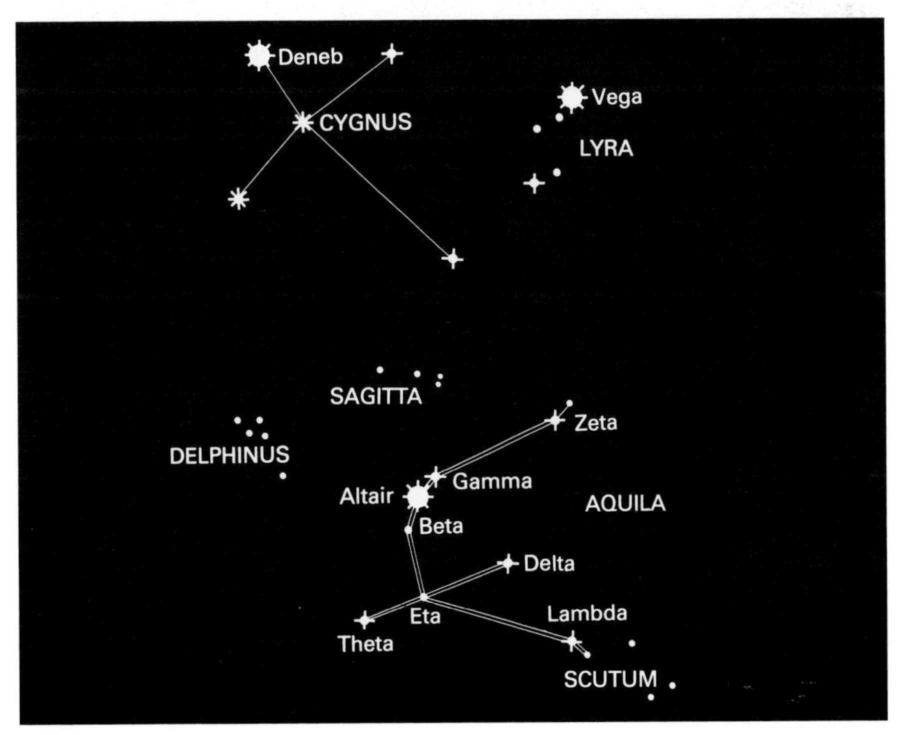

Aquila

Alkaid, in Ursa Major, may be used as a guide together with Vega, but the distance is considerable – half way across the sky – and once Vega and Deneb have been found, Altair should follow at once, particularly as it has a fainter star to either side of it.

Altair is one of the nearest of the bright stars; it is 16½ light-years away, and ten times as luminous as the Sun. It has a hot surface, and is pure white.

Aquila itself really does give a vague impression of a bird in flight. Flanking Altair are Gamma (2.7) and Beta (3.9); Gamma is orange, and with the naked eye it is clearly off-white. Zeta (3.0) lies in the northern part of the constellation.

Below the Altair trio lie three more stars, Theta (3.3), Delta (3.4) and Eta, which is a Cepheid variable with a range of 3.4 to 4.7 and a period of 7.2 days. Theta, Delta and Iota (4.3) are suitable comparison stars.

Aquila, like Cygnus, is involved in the Milky Way; the main branch goes straight through it, passing by Altair and continuing to the southernmost fairly bright star in the constellation, Lambda (3.4).

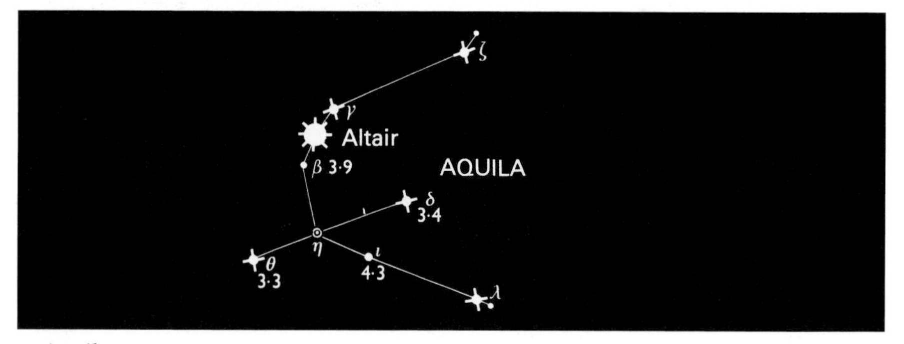

η Aquilae

SCUTUM: *the Shield*

Scutum, which contains no star above magnitude 3.8, does however contain a bright section of the Milky Way, and one cluster visible with the naked eye: M. 11, nicknamed the Wild Duck. You may see it close to Lambda Aquilae, but it appears only as a small patch, and the Milky Way nearby is so bright that identification can be troublesome. Since there is no point in giving a separate chart for Scutum, I have included it with Aquila.

DELPHINUS: *the Dolphin*

The area enclosed by lines joining Deneb, Vega, Altair and Epsilon Pegasi contains four small constellations. Epsilon Pegasi does not belong to the Square (see p. 95); it is slightly below the second magnitude, and during July evenings it is well above the eastern horizon.

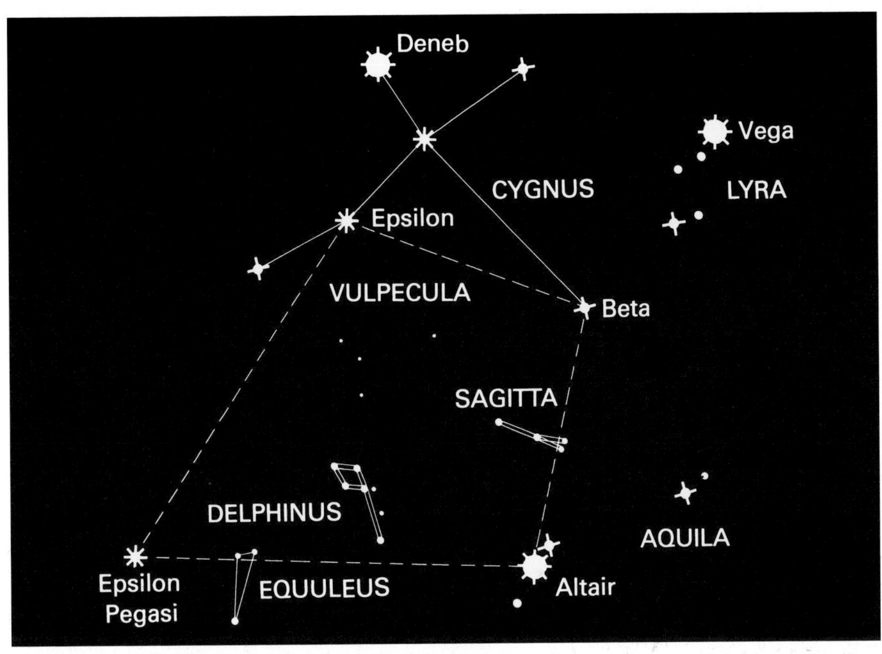

Equuleus, Delphinus, Sagitta and Vulpecula

Delphinus is easily recognized. Its brightest stars are only of magnitude 3.5, but there is a compact little group of them, appearing like a very loose cluster; unwary observers have even confused it with the Pleiades. A line from Deneb passed through Epsilon Cygni, in the X, and continued for some way on the far side will arrive near Delphinus, though the alignment is not particularly good. On the whole this is a confusing area; nothing seems to point to anything.

EQUULEUS: *the Little Horse*

Equuleus, though an ancient constellation, is very obscure. Its stars are fainter than those of Delphinus, and much less compact. Look slightly below a line joining Delphinus to Epsilon Pegasi and you will be able to see the faint triangle marking Equuleus, but there is nothing here of interest.

SAGITTA: *the Arrow*

This is another ancient constellation. Only two of its stars are above the fourth magnitude, but at least there is a distinctive shape, and it is not too difficult to conjure up the picture of an arrow. There is no serious danger of confusing Sagitta with Delphinus, because Sagitta lies almost midway between Altair and Beta Cygni, and in any case it does not give the same cluster-like impression. The Milky Way runs straight through it.

VULPECULA: *the Fox*

Vulpecula was recorded by Hevelius, though with no good reason. It lies between Sagitta and Delphinus to one side and Cygnus to the other, but is difficult to identify, because the brightest star in it is of only magnitude 4.4.

HERCULES

Hercules is a large constellation, but not a bright one, and the key to this whole area is Alpha Ophiuchi, often called by its proper name of Rasalhague. Hercules lies mainly between Vega and Arcturus – or, more precisely, between Vega and the leader of the Northern Crown, Alphekka or Alpha Coronae. The two brightest stars, Beta and Zeta, are only of magnitude 2.8. The shape of the constellation is so vague that it may be interpreted in several different ways. However, it contains two objects of real interest.

The first is the globular cluster M. 13, the brightest specimen of its class visible from Britain or the United States. It is 22,500 light-years away, and telescopes show it to be a magnificent symmetrical system containing hundreds of thousands of stars. It lies between Zeta and Eta (3.5), closer to Eta. It is on the fringe of naked-eye visibility, but dark, clear skies are needed if it is to be found.

The second notable object is the semi-regular variable Alpha Herculis, or Rasalgethi. It lies close to Rasalhague in Ophiuchus, but is considerably

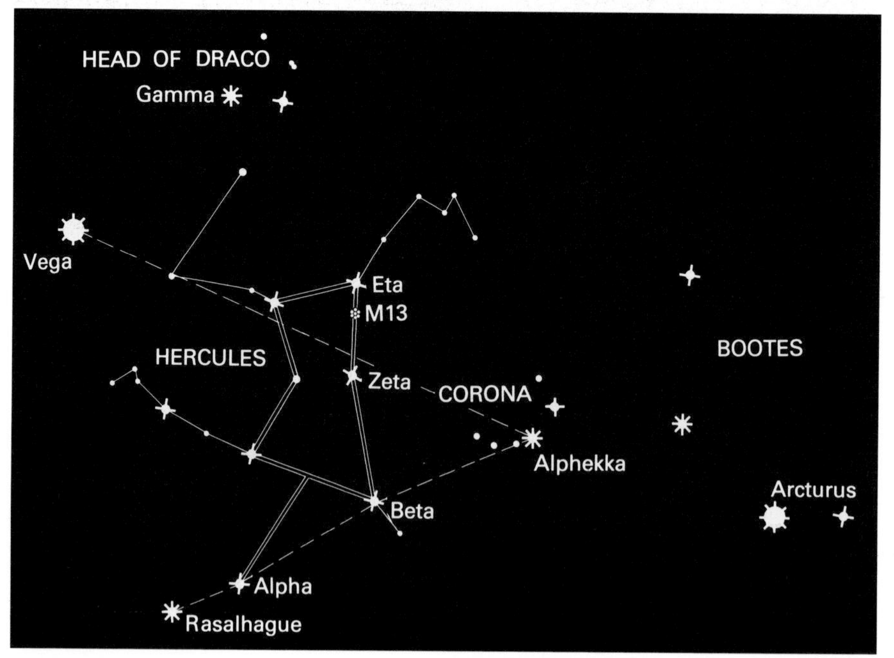

Hercules

fainter; the range of magnitude is from 3 to 4. Suitable comparison stars are Beta Herculis (2.8; rather too bright), Delta Herculis (3.1), Kappa Ophiuchi (3.2), Gamma Herculis (3.8) and Lambda Ophiuchi (3.8). Alpha Herculis is exceptionally large, with a luminosity 700 times that of the Sun.

Some time ago I pointed out Alpha Herculis to a dozen teenage beginners in astronomy. I asked them to look at it carefully and tell me if they noticed anything unusual about it. None of them realized that it was coloured, and even when I had told them, only one boy said that he could see the orange-red tint. With the telescope, of course, it was a very different story, and they all agreed that the colour could not possibly be overlooked. As I have stressed earlier, colour in a star can be seen only if the star is bright enough. All faint stars look white.

OPHIUCHUS: *the Serpent-Bearer*

Ophiuchus contains several brightish stars. As well as Alpha or Rasalhague (2.1) there are several more above magnitude 3, but there is no well-marked pattern, and most of the constellation is decidedly barren.

Rasalhague makes up a sort of kite-shape with Vega, Altair and Deneb, but probably the best way to find it is to look directly between Altair and Arcturus, rather closer to Altair. It is one of a triangle of stars, the others being Kappa (3.2) and the red variable Alpha Herculis.

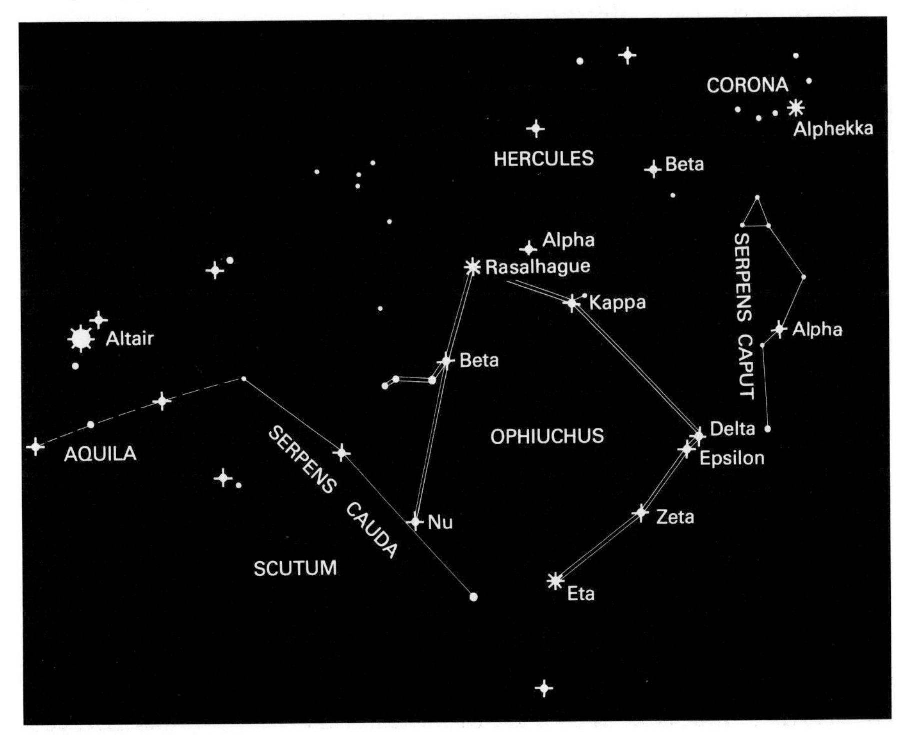

Serpens and Ophiuchus

There are two other parts of Ophiuchus recognizable without difficulty. The little group of five stars below Rasalhague, of which Beta (2.8) is the brightest, is quite conspicuous. Further south lies a curved line of four stars, of which Delta (2.7) and Epsilon (3.2) make up a noticeable pair. Ophiuchus covers a large area, but contains little of interest to the naked-eye observer, though it crosses the Zodiac and planets may therefore be found in it.

SERPENS: *the Serpent*

Serpens is a peculiar constellation, since it consists of two separate parts. It is meant to represent the snake with which Ophiuchus struggled – and which seems to have had the worst of the argument, since it has been pulled in half. The head (Serpens Caput) lies to one side of Ophiuchus, and the body (Serpens Cauda) to the other.

The head section is the more conspicuous. It begins with a small triangle of stars below Corona Borealis, and from here a line extends southward, including Alpha (2.6); Alpha more or less continues the lower curve of Ophiuchus. Cauda, which lies between Ophiuchus and Aquila, is entirely unremarkable, as it contains no bright star. For obvious reasons, I have charted Serpens together with Ophiuchus.

Unravelling this dim, chaotic area takes a good deal of patience. On the July chart, it is practically due south.

SCORPIUS: *the Scorpion*

Look low in the south during a late evening in July, and you cannot fail to notice a star which is not only very bright but also very red. This is Antares, in Scorpius. Its name means 'the Rival of Mars' – the Greek name for Mars was Ares – and it is much the ruddiest of the first-magnitude stars. It is 7500 times as luminous as the Sun, and over 300 light-years away.

Antares lies well away from any of our useful pointers, but this does not much matter, as there is no star of comparable brilliance anywhere near it. Deneb and Vega give a rough line to it, though this is distorted on the hemispherical charts. Like Altair, Antares is flanked by two fainter stars.

From Britain or New York, Antares never rises high above the horizon. This is also true of the rest of the Scorpion, which is a magnificent constellation when properly visible. Since it has a distinctive outline and is made up of a string of conspicuous stars, it is extremely easy to recognize as soon as Antares itself has been found.

The Scorpion's head consists of several brightish stars, arranged in a line running roughly north to south, of which Delta (2.3) and Beta (2.6) are the most conspicuous. The British horizon cuts off part of the constellation, and the brilliant 'sting', of which Lambda or Shaula is of magnitude 1.6 and

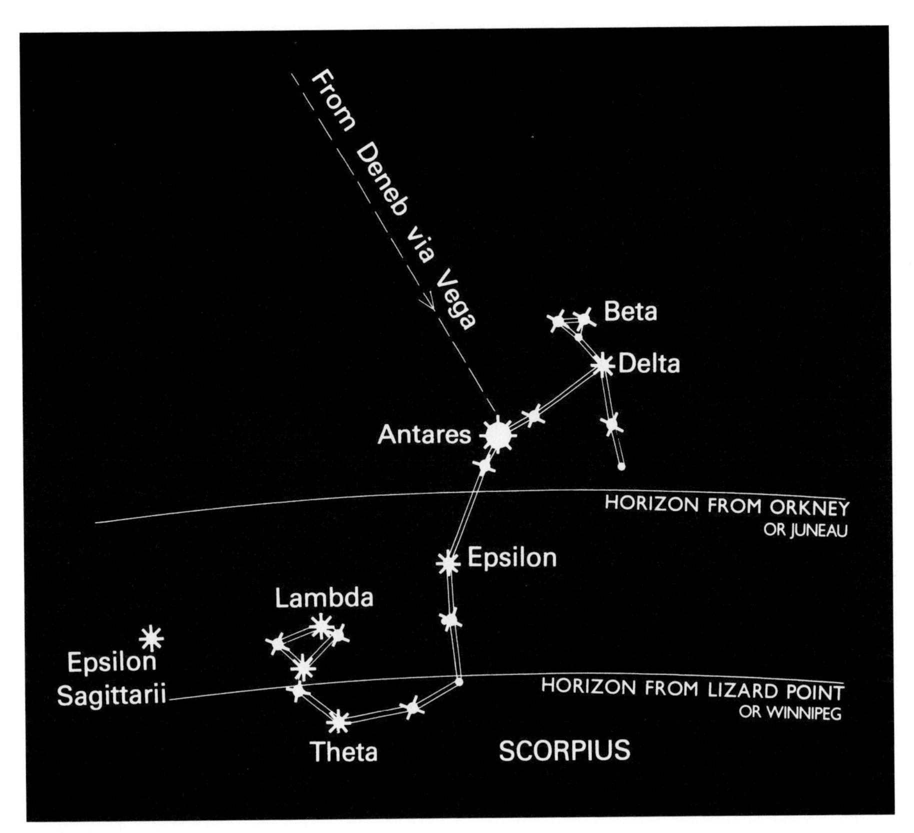

Scorpius

therefore equal to Castor, barely rises even from the southernmost part of England. From the southern United States all of the Scorpion can be seen, even to Theta (1.9). The Milky Way runs through it, and is very rich.

The constellation may be called either Scorpius or Scorpio. Astrologers usually prefer the second alternative, astronomers the first.

SAGITTARIUS: *the Archer*

Next to Scorpius lies Sagittarius, which is also in the Zodiac and is crossed by the Milky Way. It has a number of bright stars, but there is no really obvious pattern; some people have likened it to a teapot. If Antares can be seen, simply look eastward along the horizon. During July evenings this is straightforward enough, but if Antares has set the only solution is to use Deneb and Altair as guides. Continue the direction-line down towards the horizon, and you will come to Sagittarius.

The northern part of the group contains one bright star, Sigma (2.0), and several others of above the third magnitude. Further south, and rising in New York but to all intents and purposes invisible from Britain, is the leader of the constellation, Epsilon (1.8). But the main glory of Sagittarius lies in the

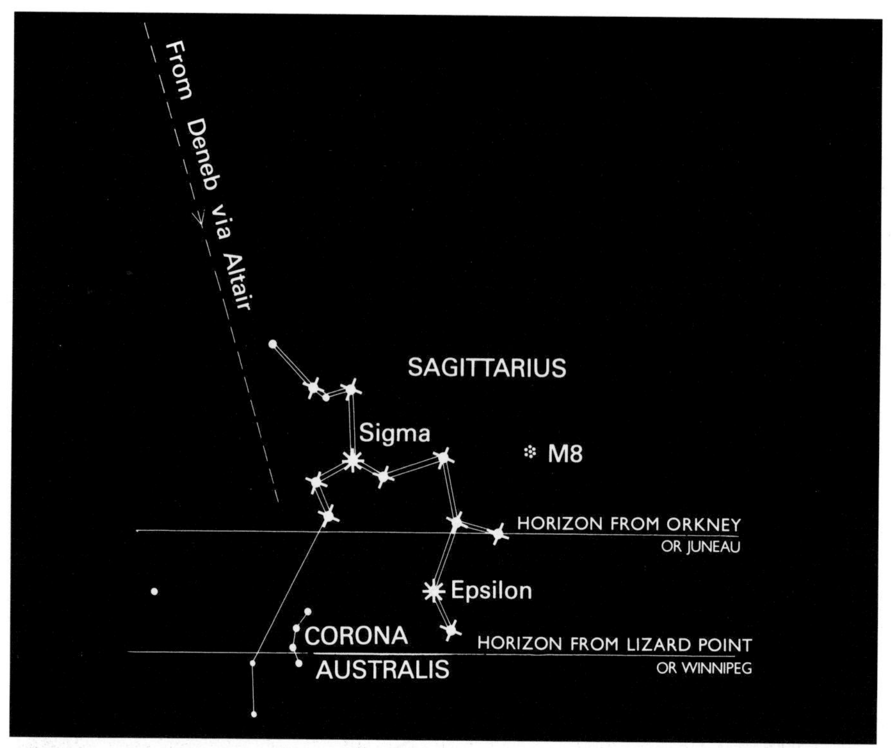

Sagittarius and Corona Australis

star-clouds which lie between us and the centre of the Galaxy. Even though they are so low down, they are striking, and binoculars show them splendidly. Keen-eyed observers may also be able to make out the misty patch which marks the Lagoon Nebula, M. 8. It is a tremendous gas-and-dust cloud of the same type as the Orion Nebula, though from Britain it is too near the horizon to be seen easily without optical aid.

CORONA AUSTRALIS: *the Southern Crown*

This is a small constellation, made up of a curved line of faint stars. I have shown it together with Sagittarius, though it is so far south that British observers will not find it; from the southern United States it will be seen easily. It contains nothing of note.

LIBRA: *the Balance*

The last of our summer groups is Libra, which is a Zodiacal constellation but is otherwise undistinguished. It is at its highest during July evenings, but is too faint and formless to be readily identifiable.

One way of finding it is to continue the main curve of the Scorpion, as shown in the diagram below. After a while you will reach Alpha Librae (2.7), after which you should be able to recognize the rest of the group, made up of a quadrilateral of stars. Beta is of magnitude 2.6, Gamma 3.9 and Sigma 3.3. Sigma was formerly included in Scorpius, as Gamma Scorpii – a reminder that in its original form Libra represented the Scorpion's claws.

Beta Librae is said to be the only star to show a perceptible greenish hue to the naked eye. I admit that I have never been able to see this myself, but certainly Beta is worth more than a casual glance. To find it, look between Antares and Arcturus, rather closer to Antares and somewhat west of the connecting line. The method is rough and ready, but it should prove adequate.

By mid-July the evenings are starting to draw in, and this is probably the best time to look for Scorpius and the other bright southern groups. They are full of interest, and more than compensate for the barrenness of Hercules and his neighbours.

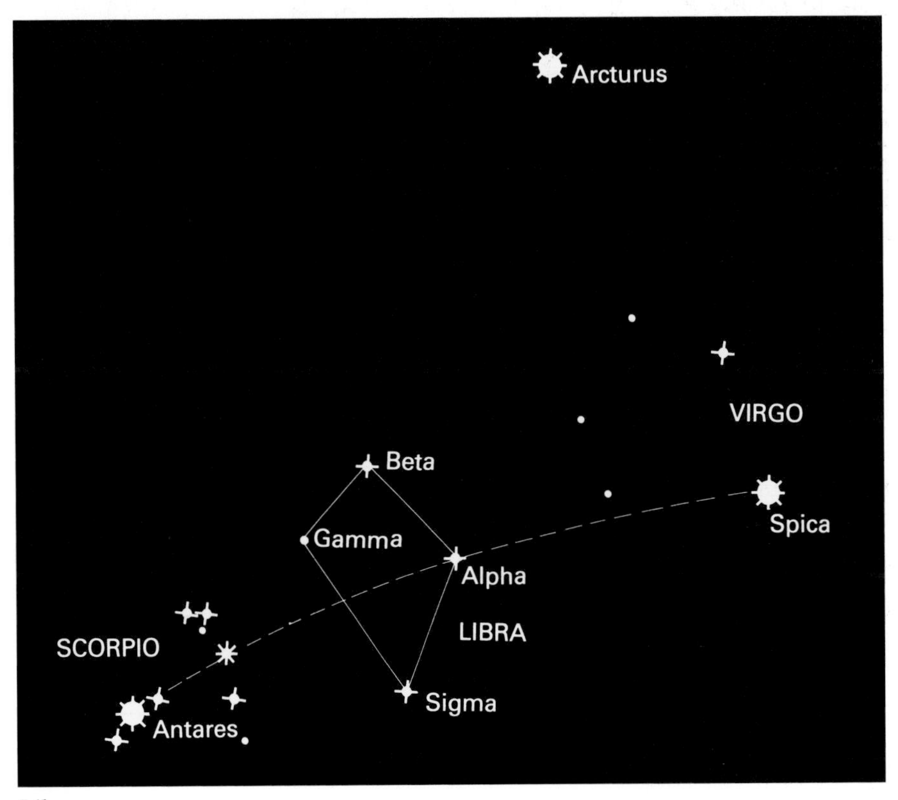

Libra

14

Northern stars: the August sky

August can be a spectacular month so far as the night sky is concerned. It marks the main meteor shower of the year, and during the first fortnight the Perseid shooting-stars are very much in evidence; I will have more to say about them later.

Meanwhile, the August charts hold good for:

1 May:	4 a.m.	(4 hours G.M.T.)
1 June:	2 a.m.	(2 hours G.M.T.)
1 July:	midnight	(0 hours G.M.T.)
1 August:	10 p.m.	(22 hours G.M.T.)
1 September:	8 p.m.	(20 hours G.M.T.)
1 October:	6 p.m.	(18 hours G.M.T.)

Ursa Major lies north-west, with Cassiopeia at about the same height in the north-east. Capella is still very low as seen from Britain or Canada, but Perseus is coming into view. The main addition to the hemispherical charts is the Square of Pegasus, characteristic of the autumn – still low, but becoming prominent during early mornings.

Pegasus will provide our autumn 'anchor'. From the Square, the line of stars marking Andromeda extends down to Perseus, and thence to Capella.

Vega is still very high up, and the Summer Triangle remains dominant. Arcturus is dropping in the west, while the formless Hercules, Ophiuchus and Serpens take up much of the southern aspect. Antares is past its best, and Libra is setting, though the stars of Sagittarius still make a reasonable display above the southern horizon.

Some more large, dull constellations are coming into view below Aquila and Pegasus. In particular there are Capricornus (the Sea-Goat) and Aquarius (the Water-Bearer), both of which lie in the Zodiac but have little else to recommend them.

With darker nights there is more chance of seeing the Milky Way, and during August evenings the luminous band is high up, passing from Capella through Perseus, Cassiopeia, Cygnus, Aquila and down to Sagittarius. The branching of the Milky Way in Cygnus is very obvious.

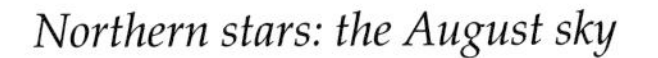

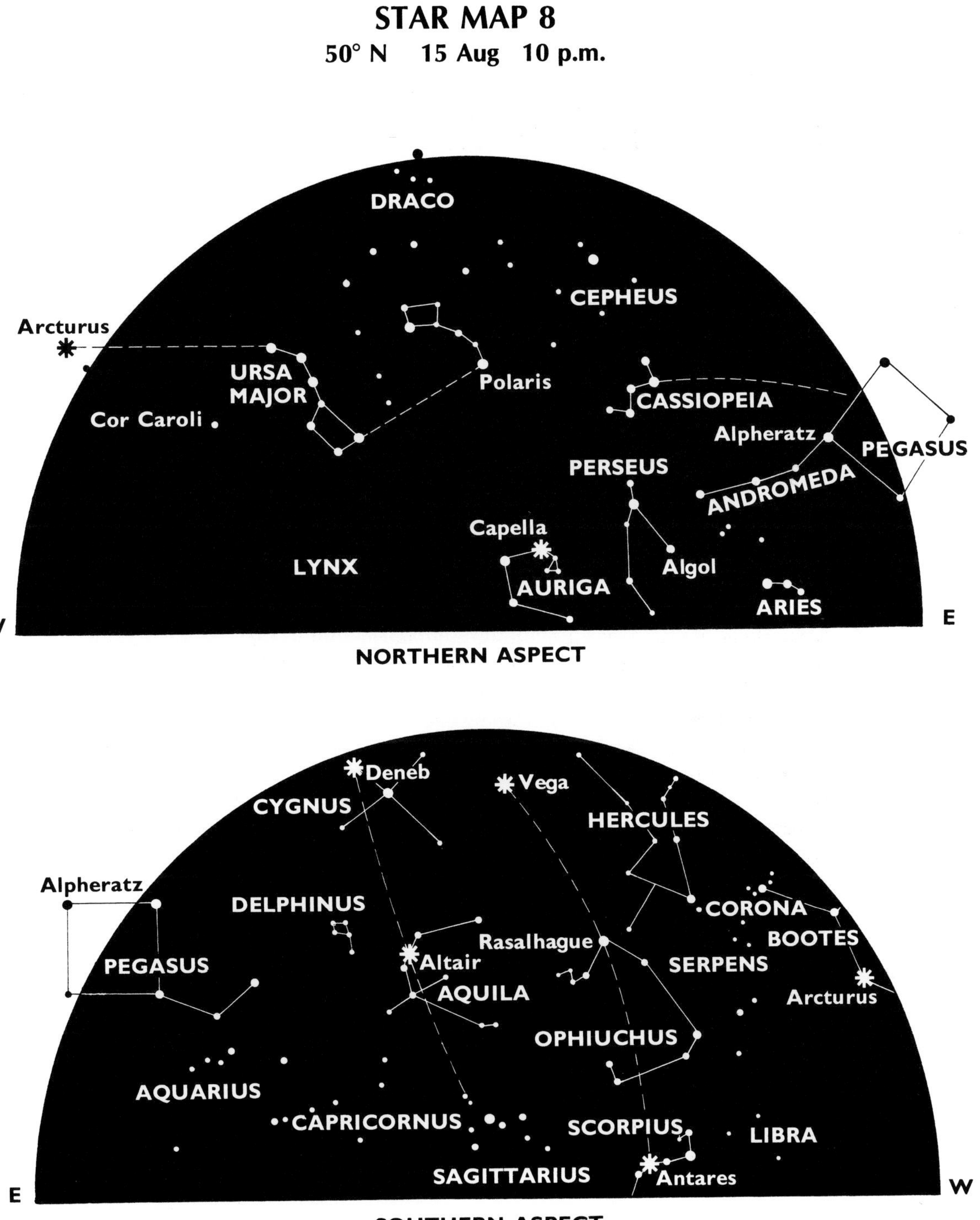
STAR MAP 8
50° N 15 Aug 10 p.m.
DRACO
CEPHEUS
Arcturus
URSA MAJOR
Polaris
Cor Caroli
CASSIOPEIA
Alpheratz
PEGASUS
PERSEUS
ANDROMEDA
Capella
LYNX
Algol
AURIGA
ARIES
E
NORTHERN ASPECT
Deneb
Vega
CYGNUS
HERCULES
Alpheratz
DELPHINUS
CORONA
BOOTES
Rasalhague
Altair
SERPENS
PEGASUS
AQUILA
Arcturus
OPHIUCHUS
AQUARIUS
CAPRICORNUS
SCORPIUS
LIBRA
SAGITTARIUS
Antares
E
W
SOUTHERN ASPECT

15

Northern stars: the September sky

The charts for September evenings show one significant feature The Pleiades have come into view in the east. I always feel that the appearance of the Pleiades before midnight marks the approach of autumn, and that the period of snowstorms, icy roads and fogs lies not far ahead.

The charts apply to the following times:

1 June:	4 a.m.	(4 hours G.M.T.)
1 July:	2 a.m.	(2 hours G.M.T.)
1 August:	midnight	(0 hours G.M.T.)
1 September:	10 p.m.	(22 hours G.M.T.)
1 October:	8 p.m.	(20 hours G.M.T.)
1 November:	6 p.m.	(18 hours G.M.T.)

The Square of Pegasus is high in the south-east, and may be found by using Cassiopeia as a pointer, though the hemispherical maps do not show this well. Deneb is almost overhead, but Vega is beginning to drop westward. Altair remains prominent, but Antares has set. Part of Sagittarius can still be seen, and Ophiuchus and Serpens take up much of the area near the western horizon, though their relative dimness means that any appreciable mist will hide them. Hercules is no brighter, but at least it is higher up.

One other first-magnitude star is on view: Fomalhaut in the Southern Fish, below the Square of Pegasus. It is as bright as Pollux, but is always very low as seen from Britain or Canada, and from northern Scotland it will probably not be seen at all.

STAR MAP 9

50° N 15 Sep 10 p.m.

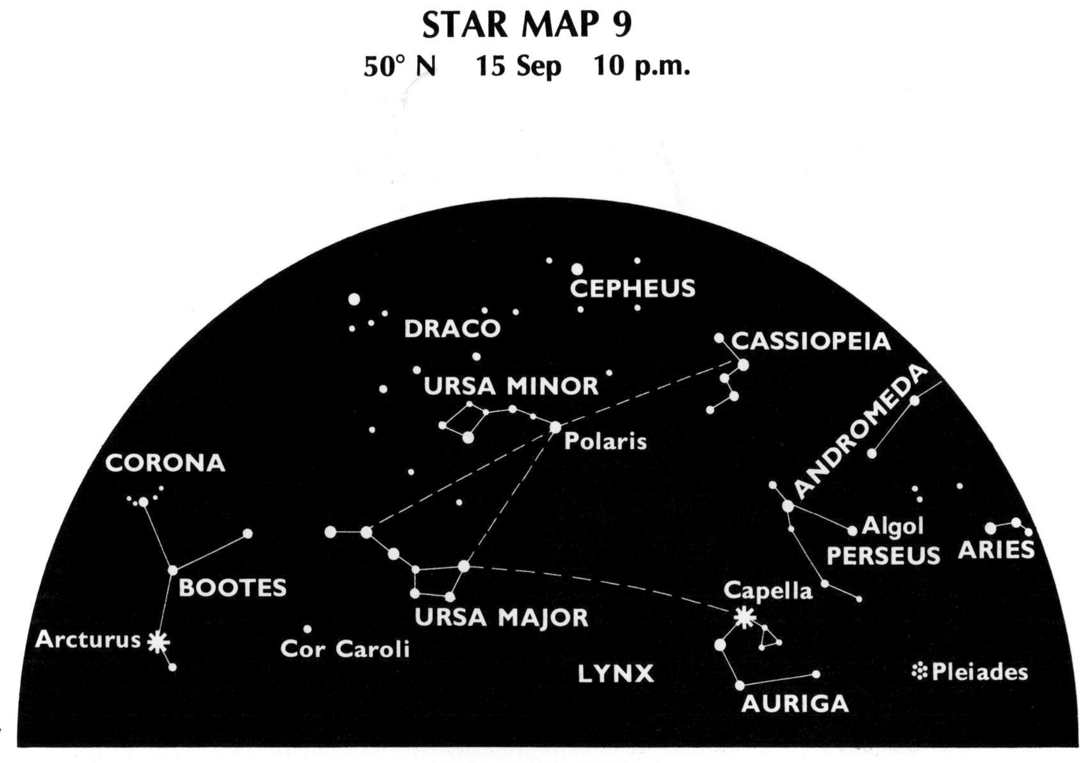

NORTHERN ASPECT

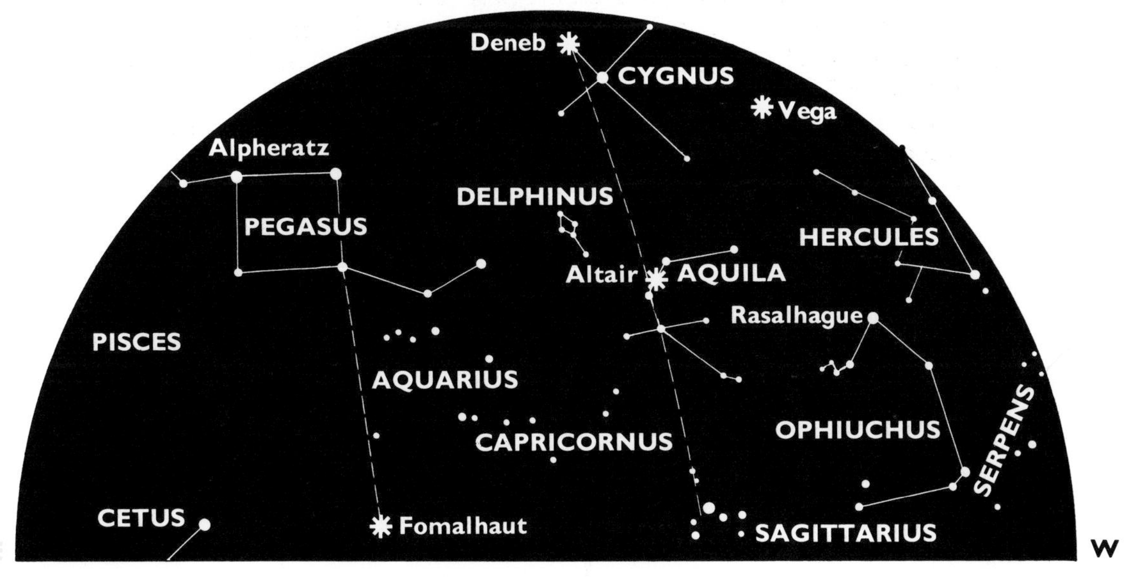

SOUTHERN ASPECT

16

Northern stars: the October sky

During October evenings, the autumn constellations are at their best. The maps are valid for:

1 July:	4 a.m.	(4 hours G.M.T.)
1 August:	2 a.m.	(2 hours G.M.T.)
1 September:	midnight	(0 hours G.M.T.)
1 October:	10 p.m.	(22 hours G.M.T.)
1 November:	8 p.m.	(20 hours G.M.T.)
1 December:	6 p.m.	(18 hours G.M.T.)

Ursa Major is low in the north; Cassiopeia is not far from the zenith, while Capella in the north-east is not a great deal lower than Vega in the north-west. Arcturus has disappeared; during October it sets early in the evening.

We have also lost part of Ophiuchus, though Hercules is still above the horizon. To replace Ophiuchus, another large, dim group – Cetus, the Whale or Sea-Monster – has appeared in the south-east. The Summer Triangle remains prominent, with Deneb very high up. The Milky Way is finely displayed, and passes overhead.

The southern aspect is dominated by Pegasus, which has appeared on our charts before.

STAR MAP 10

50° N 15 Oct 10 p.m.

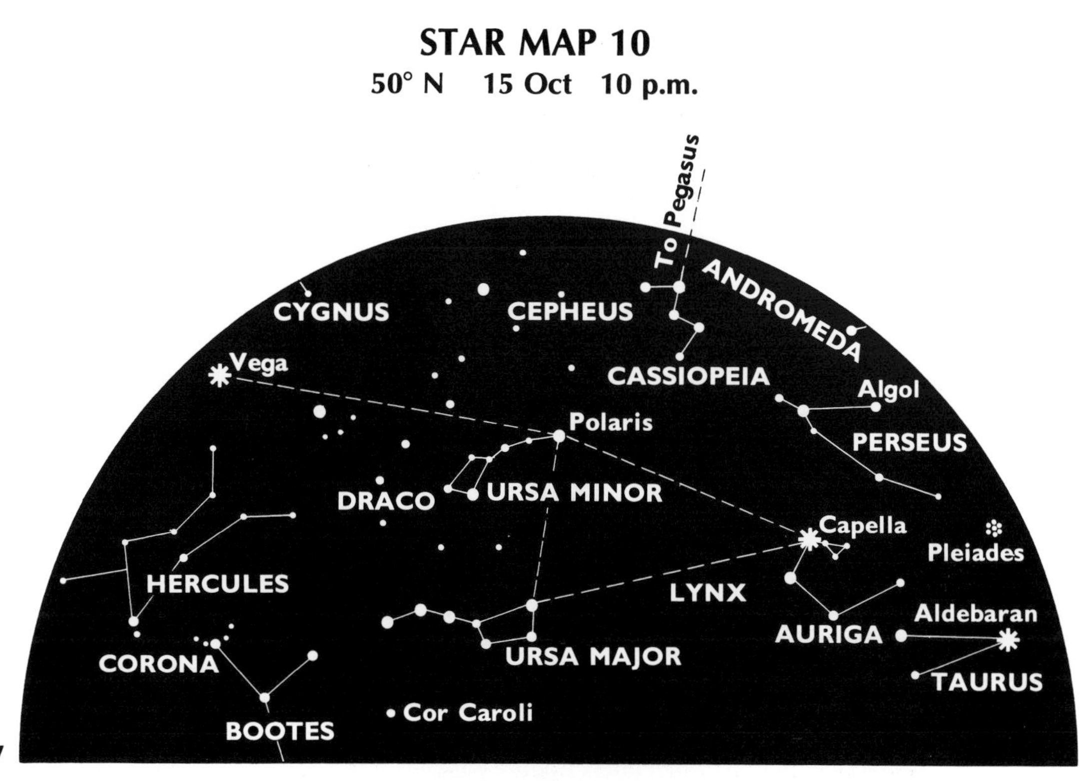

NORTHERN ASPECT

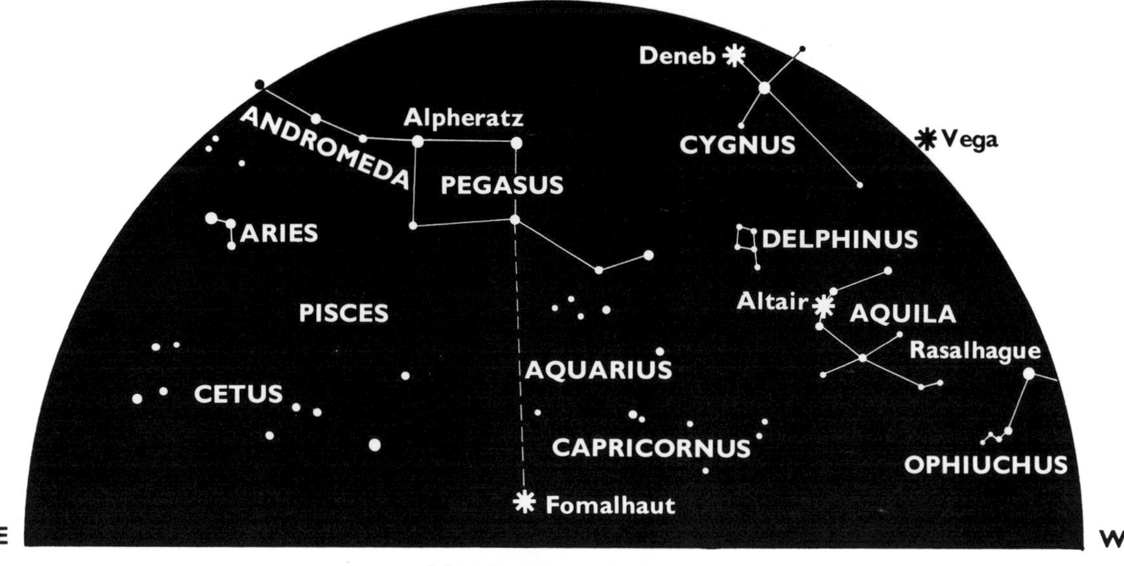

SOUTHERN ASPECT

PEGASUS: *the Flying Horse*

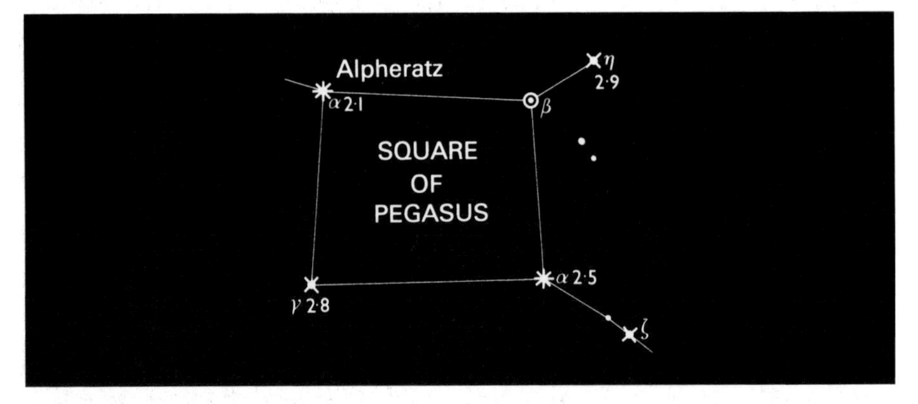

β Pegasi

On maps, Pegasus looks extremely conspicuous. The four main stars make up a square, and it might be thought that they would stand out at once. Actually, they do not – partly because they are not really bright, and partly because the Square is very large. Once found, it will easily be located again, but at the outset you will have to look for it carefully.

The W of Cassiopeia makes a good pointer; Gamma and Alpha act as guides. A line passed through them, and continued for some distance, will run straight into Pegasus. Also, look for the combined arrangement of Pegasus, Andromeda, Alpha Persei and Capella, which is not unlike an enlarged and very distorted version of Ursa Major.

The stars in the Square are Alpha Pegasi (2.5), Beta (variable), Gamma (2.8) and Alpha Andromedae (2.1). Alpha Andromedae, or Alpheratz, used to be known as Delta Pegasi. The free transfer to Andromeda seems illogical, since the star so clearly belongs to the Pegasus pattern. Gamma, the lower left-hand star, is the faintest of the four, and when there is appreciable low-lying haze it may well be hidden, thereby giving the Square an incomplete appearance. However, this will not usually be the case during October evenings, when the Square is high in the sky and well clear of horizon mist.

The most notable star in the constellation is Beta or Scheat, a vast orange-red giant whose colour is just about detectable with the naked eye. It varies between magnitudes 2.4 and 2.8, in a rough period of about 35 days. Alpha and Gamma make good comparison stars, but remember to allow for extinction; seen from northern latitudes, Beta is always much higher than Gamma or Alpha.

The Square provides a good demonstration of how uncrowded the sky really is. The area covered is large, but not many faint stars can be seen in it with the naked eye; anyone who counts a dozen will be doing well.

The only other important star in Pegasus is Epsilon (2.4), which is well away from the Square, in a relatively isolated position between Alpha Pegasi and Altair. It has been suspected of slight variability, and is worth watching. Binoculars show that it is decidedly orange.

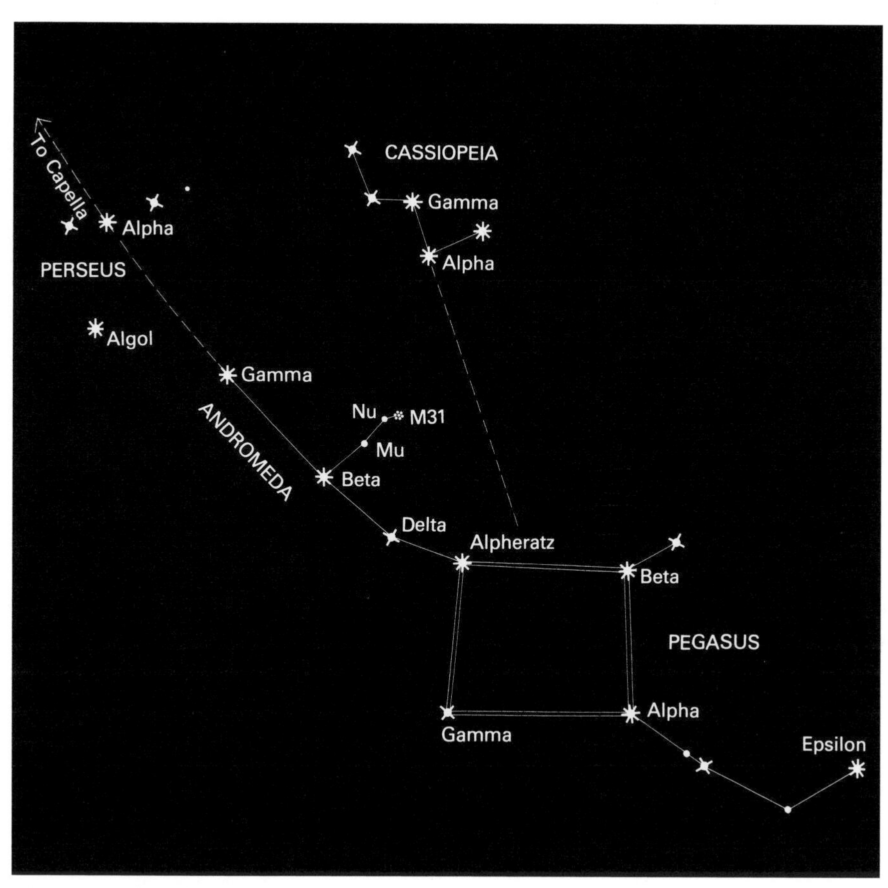

Pegasus and Andromeda

ANDROMEDA

Andromeda joins on to Pegasus; as we have seen, Alpheratz or Alpha Andromedae is in the Square. The other principal stars are, in order, Delta (3.3), Beta (2.1) and Gamma (also 2.1). There is an interesting colour-contrast here. Alpheratz is white and Gamma slightly orange, while Beta is definitely orange-red. In fact Beta is the same colour as Betelgeux, though since the star is more than a magnitude fainter the hue is not obvious.

The main object of interest is M. 31, the Great Spiral, which is the brightest of the external galaxies visible from Britain or the United States. To locate it, first identify Beta and then look upward; you will see two fainter stars, Mu (3.9) and Nu (4.5). The Spiral lies close to Nu, slightly to the right. It is just visible with the naked eye, but you will need a perfectly dark and clear sky. Through a small or even a moderate-sized telescope it is frankly disappointing; photographs taken with large instruments are needed to show the spiral form clearly.

ARIES: *the Ram*

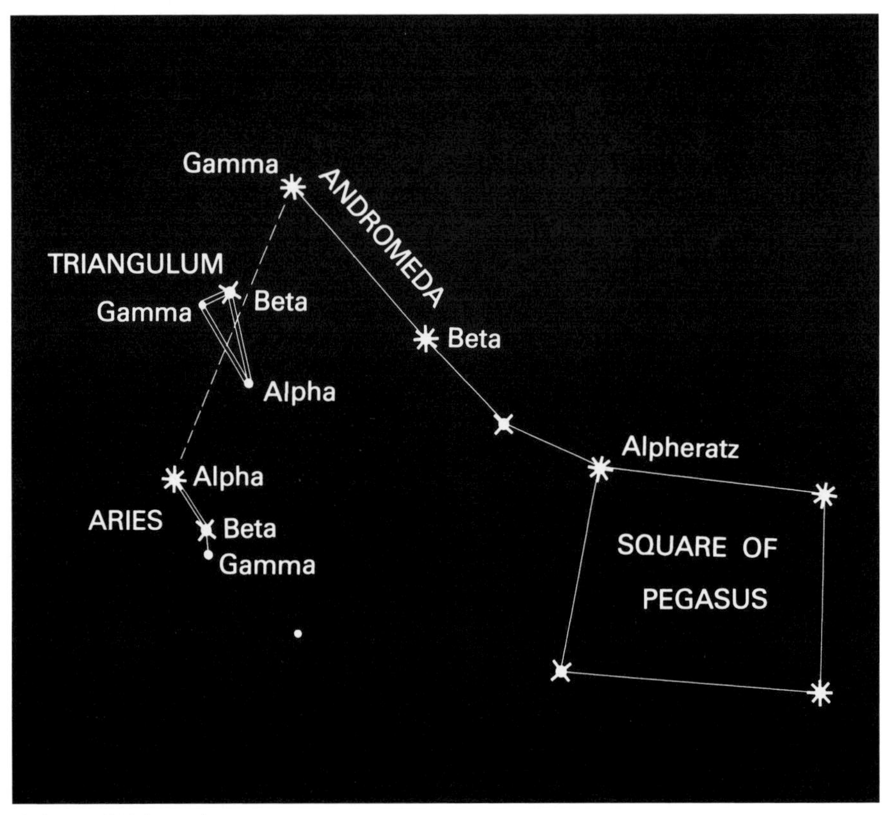

Aries and Triangulum

Alpha Arietis or Hamal (2.0), the leader of the constellation, may be found below the Andromeda chain. It is rather isolated, and is slightly orange, though to the naked eye it will probably look white. There are two more stars of reasonable brilliancy: Beta (2.6) and Gamma (3.9). Gamma is a wide double, but as it appears single with the naked eye it does not concern us here.

TRIANGULUM: *the Triangle*

As already noted, it is unusual to find a constellation which looks in the least like its namesake. Triangulum is one exception; the three main stars are Beta (3.0), Alpha (3.4) and Gamma (4.0). The group lies between Alpha Arietis and Gamma Andromedae. The most famous feature is the spiral galaxy M. 33, which is slightly further away than the Andromeda galaxy but is still a member of the Local Group. It lies rather west of a line joining Alpha Trianguli to Beta Andromedae. Some have claimed to be able just to see it with the naked eye, but I have never been able to do so without binoculars or a telescope.

PISCES: *the Fishes*

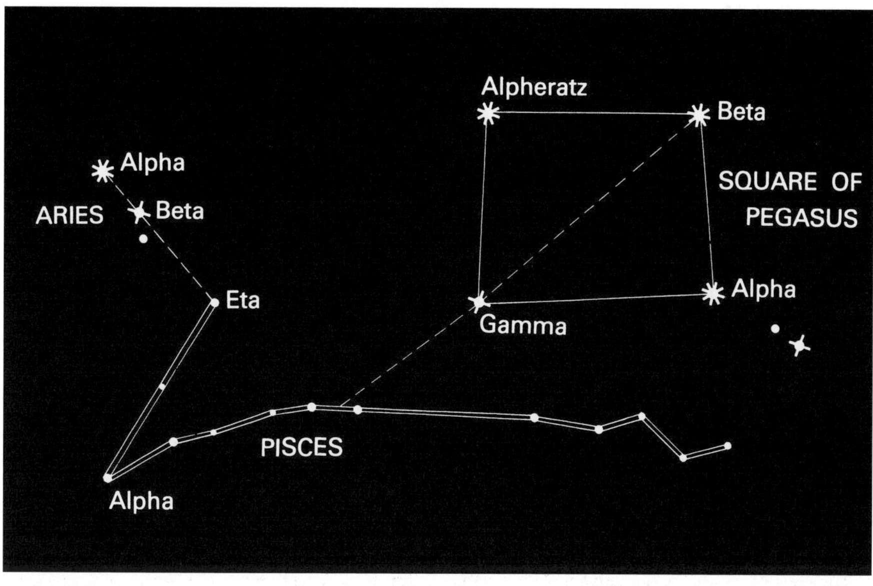

Pisces

Pisces is a Zodiacal constellation. The Sun lies in it around 22 March, when it crosses the celestial equator, moving from south to north; this is the official beginning of spring in the northern hemisphere. During October evenings Pisces is high up, but it is not easy to identify because its stars are so faint. Even the brightest of them, Eta, is only of magnitude 3.6. The best way to find it is to use Alpha and Beta Arietis as guides.

The main constellation consists of a long chain of stars below Andromeda and the Square of Pegasus. If you connect Beta and Gamma Pegasi, and continue the line for an equal distance, you will come to the middle of Pisces; you should then be able to make out the group, but there is nothing of interest to be seen. The Pisces chain ends below Alpha Pegasi in the Square.

PISCIS AUSTRINUS: *the Southern Fish*

The region near the southern horizon during October evenings is barren, but it does contain one bright star. This is Fomalhaut, in the constellation sometimes known as as Piscis Australis or Piscis Austrinus.

Fomalhaut is of magnitude 1.2, so that it is brighter than Regulus. It is never well seen from Europe or Canada, but luckily Beta and Alpha Pegasi give a perfect guide-line to it: the pointer passes across a region so dim that Fomalhaut is the first bright star to be reached. Actually, the best time to look for it is around 8 to 9 p.m. in mid-October; by midnight it will have disappeared, though from the latitude of New York it is higher up and observable for longer.

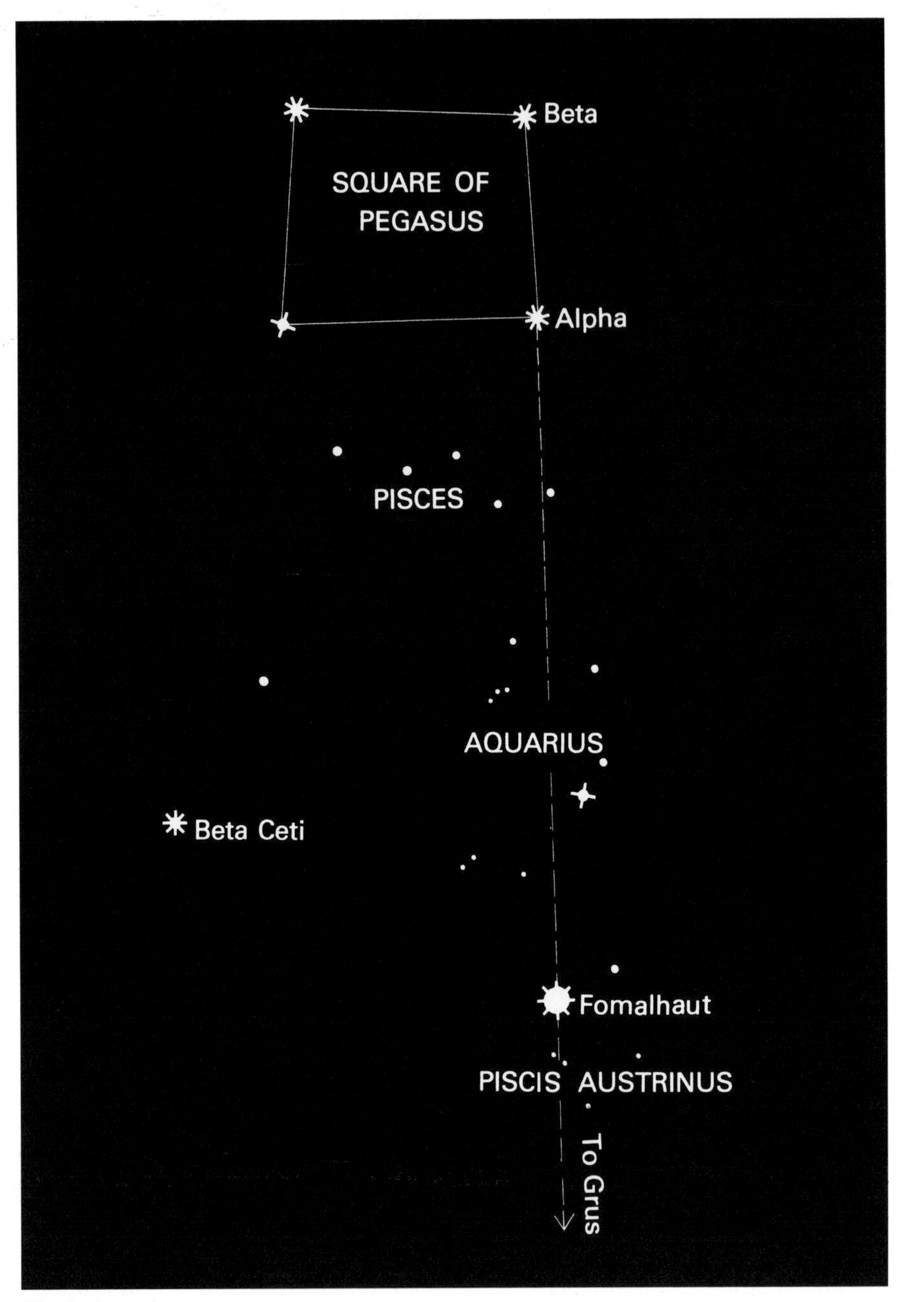

Piscis Austrinus

Fomalhaut is white. It is only 22 light-years away, and 13 times as luminous as the Sun. Like Vega, it was studied from the IRAS satellite and found to be associated with cool material which may or may not indicate the presence of a planetary system. There are no other stars in the Southern Fish above the fourth magnitude.

GRUS: *the Crane*

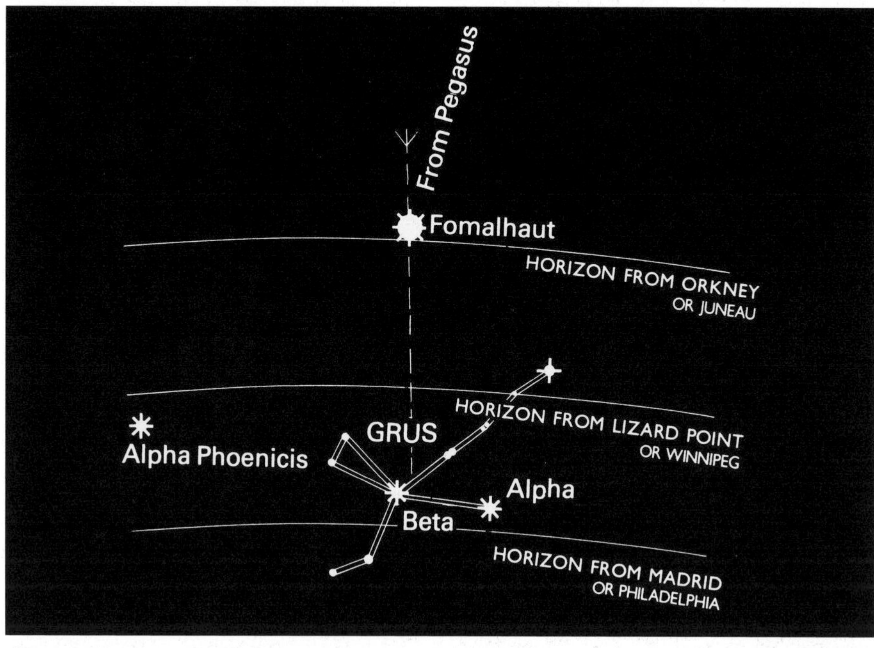

Grus

Grus, the brightest of the four 'Southern Birds' to be described later (see p. 142), is never visible from Britain or the northern United States, but from Florida or California it is quite prominent. The best pointer to use is a line starting at the Square of Pegasus and passing through Fomalhaut; this will lead to the two leaders of Grus, Alpha (Alnair) and Beta. I will have more to say about Grus when we come to consider the southern stars.

CETUS: *the Whale*

Cetus is yet another large constellation with few bright stars, but it is not hard to identify the leader, Beta or Diphda (2.0), by using Alpheratz and Gamma Pegasi as pointers. The lining-up is not exact, but it is good enough. Beware of confusing Beta Ceti with Fomalhaut. There is no real problem, because Fomalhaut is further west, considerably lower down and almost a magnitude brighter.

Having located Beta at one end of Cetus, the next step is to find the Whale's head at the other end. For this, Beta Andromedae and Alpha Arietis may be used. The head is marked by a quadrilateral of stars; Alpha or Menkar (2.5) is the brightest of them, and is orange.

Next, check the pattern of stars between Beta and Alpha, as shown here. It is worth finding Tau (3.5), which is less than 12 light-years away and is

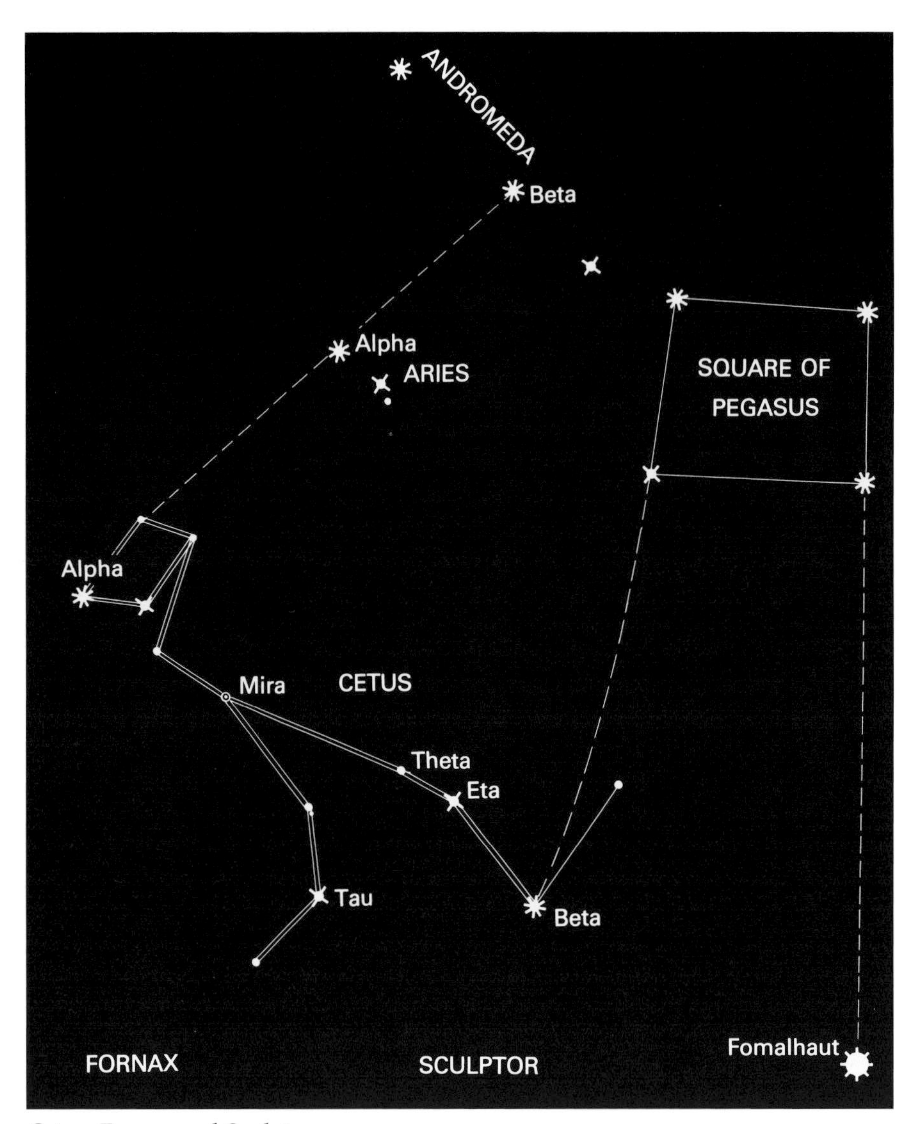

Cetus, Fornax and Sculptor

cooler and fainter than the Sun and sometimes regarded as a possible planetary centre. However, the most intriguing feature of Cetus is Omicron or Mira, the best known of the long-period variables. The mean period is 331 days. At some maxima, as for instance that of 1987, the magnitude may rise to 2, while on other occasions it is no brighter than 4. At minimum it drops to magnitude 9, far below naked-eye visibility. Suitable comparison stars are given in the chart. Mira is visible without optical aid for only a few weeks in every year, and unwary observers have often mistaken it for a nova.

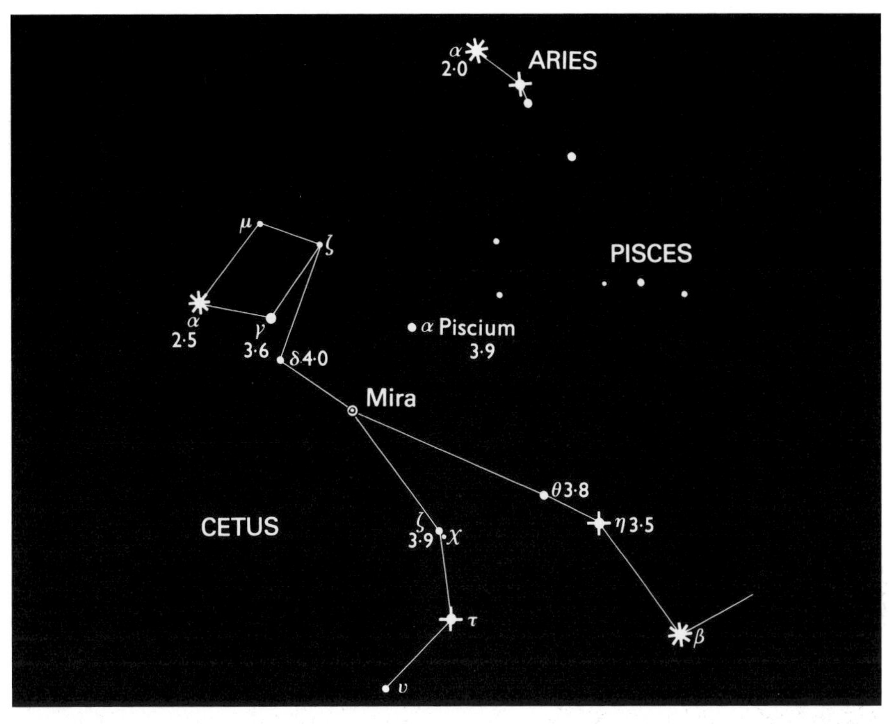

o Ceti

FORNAX: *the Furnace* and SCULPTOR: *the Sculptor*

For the sake of completeness, I must mention two very obscure groups adjoining Cetus: Fornax and Sculptor, both added to the sky by the French astronomer Lacaille in 1752. Both are very low as seen from northern countries, and both are devoid of bright stars or interesting objects, so that I have merely indicated their positions on the map given for Cetus.

CAPRICORNUS: *the Sea-Goat*

The idea of a marine goat may seem rather peculiar; in some legends the Zodiacal constellation of Capricornus is identified with the demigod Pan. There is only one notable object, the naked-eye double Alpha. The brightest star is Delta (2.9).

The best way to locate Capricornus is to start with Altair in Aquila. The line of three stars of which Altair is the central member points straight down to the Sea-Goat, arriving at Alpha, which is made up of two components of magnitudes 3.7 and 4.5. Both may be seen easily enough, since the distance between them amounts to 376 seconds of arc. Beta (3.1), which lies nearby, is also double, but the components are too close to be seen separately with the naked eye.

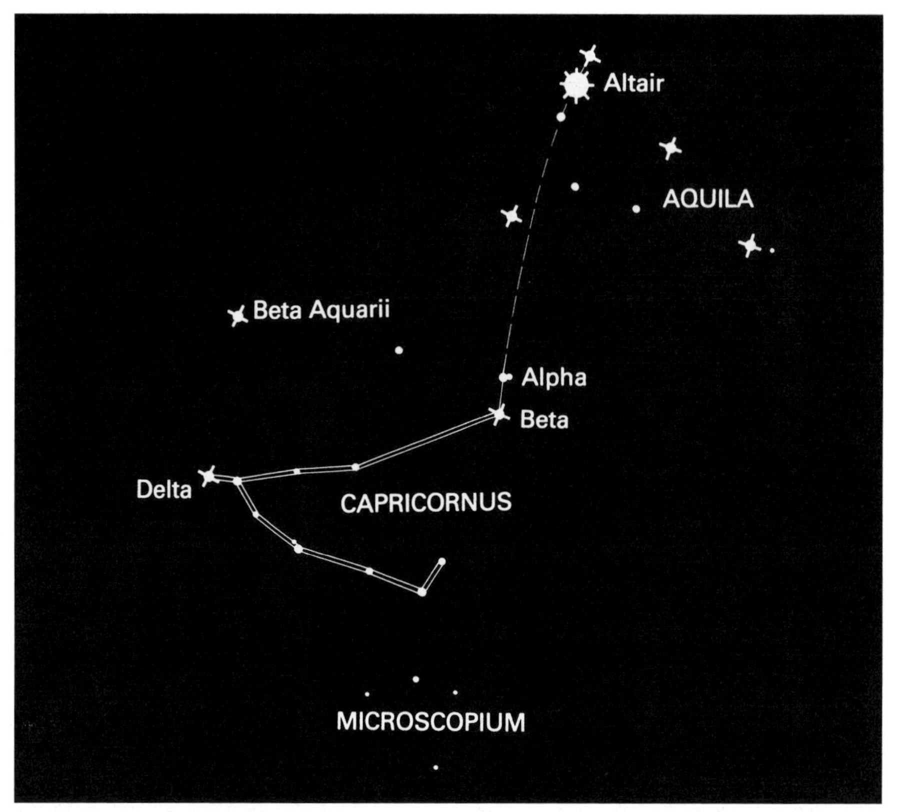

Capricornus and Microscopium

MICROSCOPIUM: *the Microscope*

Below Capricornus we come to a small, faint constellation whose modern name, Microscopium, shows that it is a late addition to the sky. Most of it rises from Britain, and all of it from New York, but it is entirely unremarkable.

AQUARIUS: *the Water-Bearer*

Finally in our description of the northern sky we come to Aquarius, the 'Man with the Watering-Pot'. We are ending on a subdued note, since although Aquarius is in the Zodiac and is one of the ancient groups, it is formless and faint.

Again the Square of Pegasus provides a guide. A line from Alpheratz, passed through Alpha Pegasi and continued for a somewhat greater distance beyond, will arrive at Alpha Aquarii (3.0), one of a quartet arranged in an irregular line. Beyond Alpha you will see Beta (2.9). The only other reasonably bright star in the constellation, Delta (3.3), lies more or less along the line joining Alpha Pegasi to Fomalhaut.

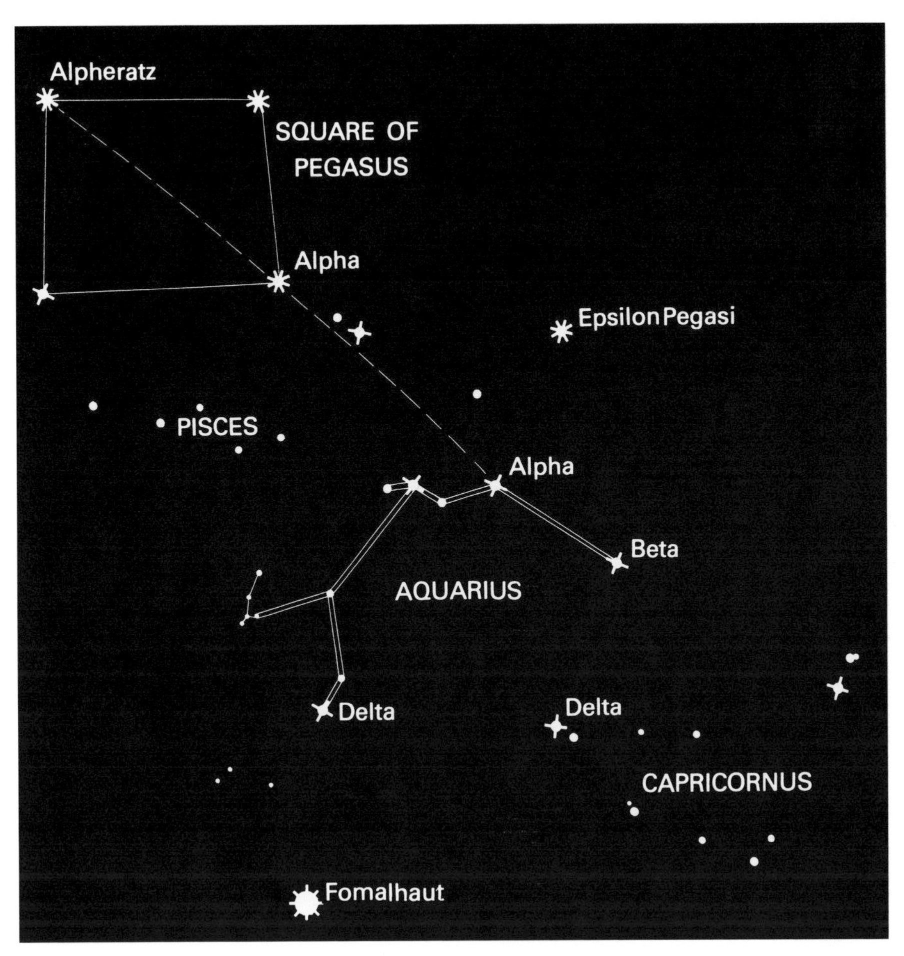

Aquarius

In point of fact Aquarius spreads over a wide area, touching Capricornus on one side and Cetus on the other. The direction-line used to find Fomalhaut, running from Beta and Alpha Pegasi, will first cross the tail of Pisces and then extend into Aquarius. Roughly half-way between Alpha Pegasi and Fomalhaut, slightly to the east of the line, may be seen a small group of stars which looks rather like a cluster, though it is not a genuine grouping and is of no real significance. It is worth looking at with binoculars, because several of its stars are orange in colour.

All in all, the area of Capricornus, Aquarius, Pisces and Cetus is one of the most barren parts of the sky. The best that can be said of it is that it is less confusing than the Ophiuchus area, and since part of it lies in the Zodiac it is sometimes graced by a brilliant planet. Should you see a really bright object there, you may be sure that it is a planet; of the stars, only Fomalhaut is bright enough to be identifiable at a glance.

17

Northern stars: the November sky

The November charts reintroduce a familiar and brilliant constellation: Orion is visible once more, though it does not come into full view much before midnight. The charts apply to:

1 August:	4 a.m.	(4 hours G.M.T.)
1 September:	2 a.m.	(2 hours G.M.T.)
1 October:	midnight	(0 hours G.M.T.)
1 November:	10 p.m.	(22 hours G.M.T.)
1 December:	8 p.m.	(20 hours G.M.T.)
1 January:	6 p.m.	(18 hours G.M.T.)

Orion is not yet high enough to be used as a general guide, but Aldebaran and the Pleiades have become prominent, while Castor and Pollux are rising in the east. The Summer Triangle has lost its dominance; Vega and Deneb are still conspicuous, but Altair has become low in the west. Capricornus has gone, and Fomalhaut grazes the horizon, though Cetus and Aquarius are still on view.

Pegasus is in the high south, somewhat west of the meridian; the Andromeda line leads to Alpha Persei, not far from the zenith. Cassiopeia is overhead; Ursa Major at its lowest. Arcturus, of course, is invisible. If you want to see Arcturus on a November night, you must wait until the early hours.

The Milky Way passes through the overhead point during the late evening, and on a clear, frosty night it is superb.

STAR MAP 11

50° N 15 Nov 10 p.m.

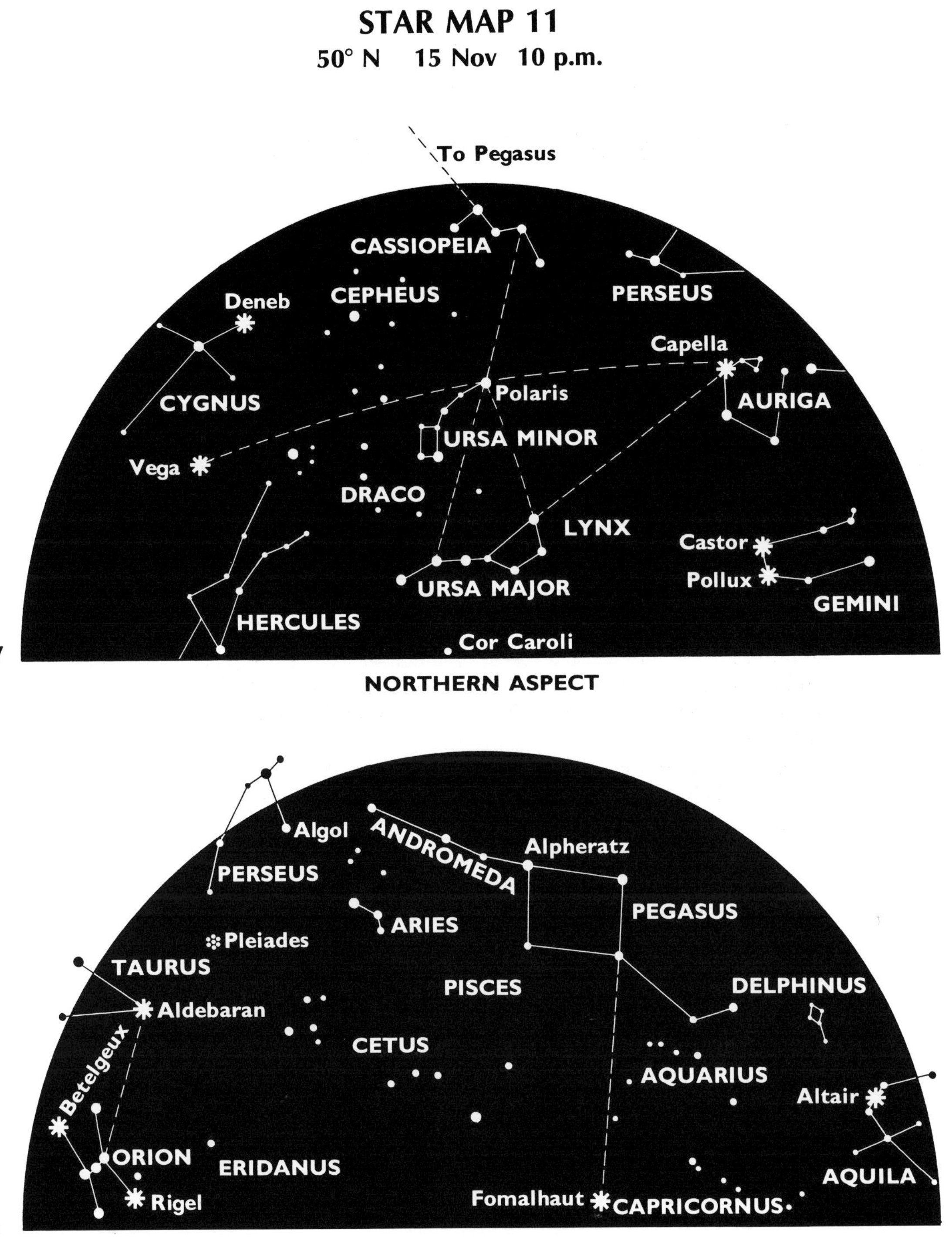

18

Northern stars: the December sky

The December charts bring us back almost to our starting-point, with Ursa Major standing on its tail and Orion visible once more in the south-east. The charts apply to:

1 September:	4 a.m.	(4 hours G.M.T.)
1 October:	2 a.m.	(2 hours G.M.T.)
1 November:	midnight	(0 hours G.M.T.)
1 December:	10 p.m.	(22 hours G.M.T.)
1 January:	8 p.m.	(20 hours G.M.T.)
1 February:	6 p.m.	(18 hours G.M.T.)

All the main guides – Orion, Ursa Major, Cassiopeia and the Square of Pegasus – are on view. Orion is not yet at its best, but it is well above the horizon. Sirius is barely visible, but Capella is high. Vega has dropped almost out of sight, and Altair has gone altogether, though Deneb is still on view. The Milky Way makes a fine showing, stretching right across the sky from Cygnus in the north-west through to Monoceros in the south-east. Leo has not fully appeared, but makes its entry in the east during the early hours of a December morning, and is high well before sunrise.

STAR MAP 12

50° N 15 Dec 10 p.m.

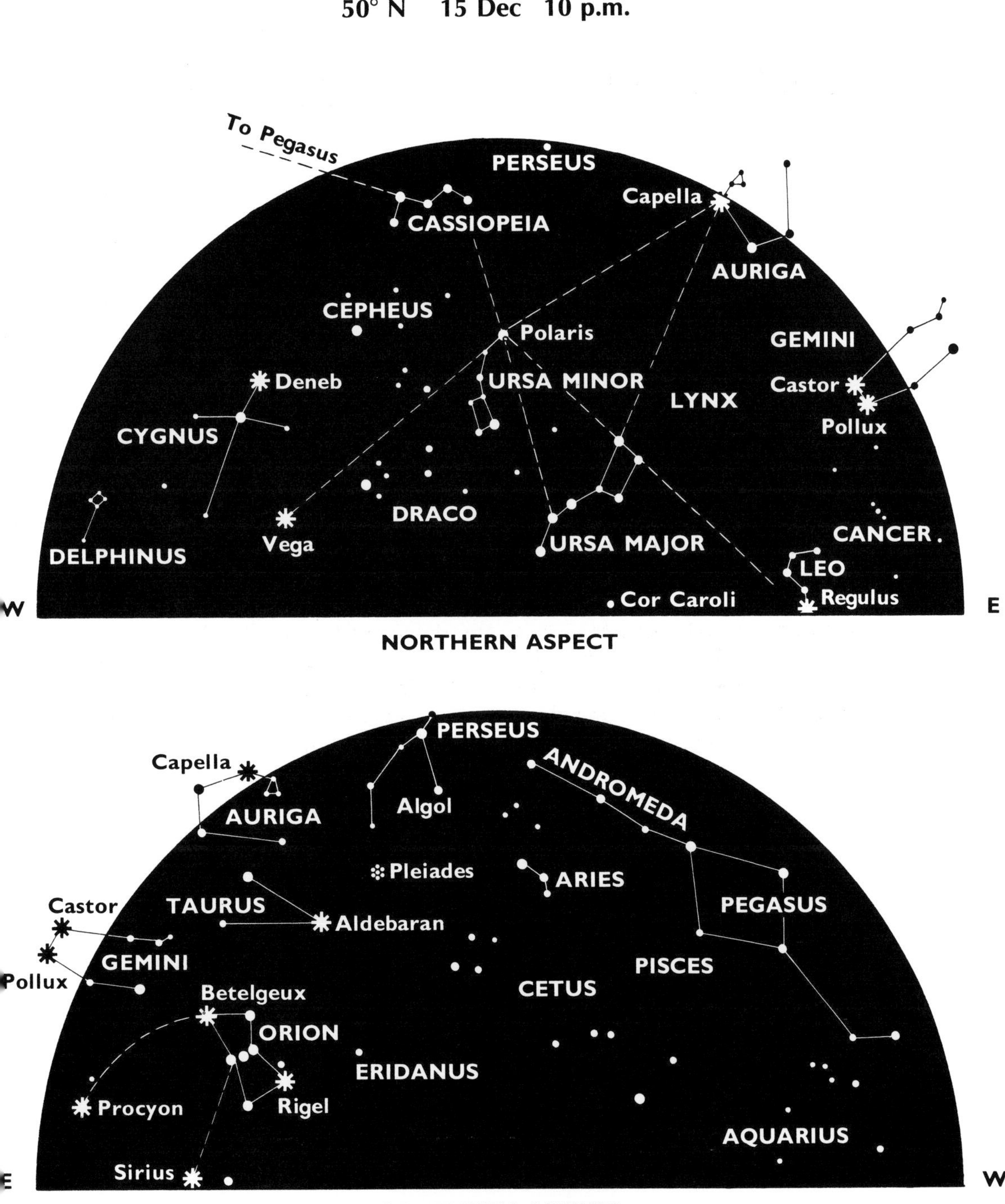

19

Stars of the far south

It was easy enough to begin a description of the northern hemisphere of the sky with the polar region, because of the presence of distinctive constellations such as the Bears and Cassiopeia. Things are much less convenient in the south, because the polar area is remarkably barren, and, as we have noted, there is no bright south pole star; Sigma Octantis is a poor substitute for the northern Polaris.

The major problem, therefore, is the lack of useful guides. Orion is invaluable when it can be seen, but during winter it is out of view. Ursa Major is useless; admittedly it reaches a respectable altitude from Darwin, and parts of it can be seen over most of Australia and South Africa, but it skirts the horizon and is never visible for long. From New Zealand it has to all intents and purposes been lost altogether, and none of the seven chief stars rises from latitudes south of Auckland.

The starting-point must be Crux Australis, the Southern Cross, which is the smallest constellation in the sky but is also one of the most brilliant. It is not a true cross; there is no X-form, as with Cygnus, and the shape is much more like that of a kite, but there must be few observers in the southern hemisphere who fail to identify it, particularly as it is close to the two brilliant southern Pointers, Alpha and Beta Centauri; so I will assume you know how to find the Cross, and work from there.

The next step is to locate the south pole, which, frankly, is not easy; except when the sky is dark and clear, and there are no artificial lights anywhere around, the whole region will appear virtually blank. The method I recommend is to follow the line of the longer axis of the Cross until you come to a really brilliant star, Achernar in Eridanus. The pole lies roughly midway between Achernar and the Cross. The trouble here is that when either of these two skymarks is at its lowest, it will probably be lost in horizon haze, and from Johannesburg both actually set; the Cross is not circumpolar anywhere north of Sydney or Cape Town. Unfortunately it is hard to give a better procedure. Beware, too, of the False Cross, which lies partly in Carina and partly in Vela. The arrangement of the four stars is much the same as with Crux, though they are not so bright, and the False Cross is much the larger of the two.

It will, I feel, be best to describe the main constellations with the seasonal maps, but we must first dispose of those which lie around the pole and are circumpolar from New Zealand and much of South America, Australia and South Africa.

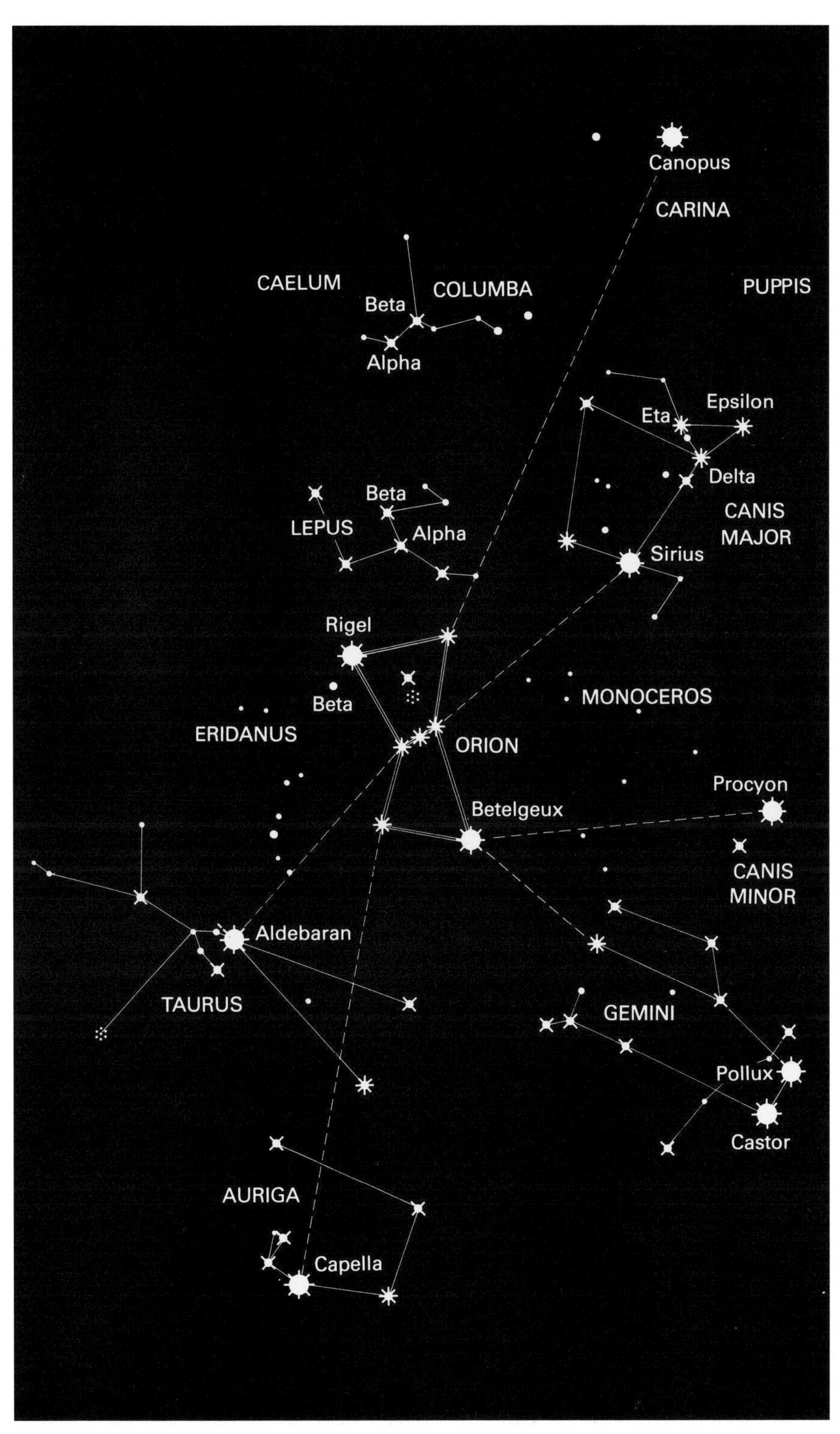

Orion: guide to the south

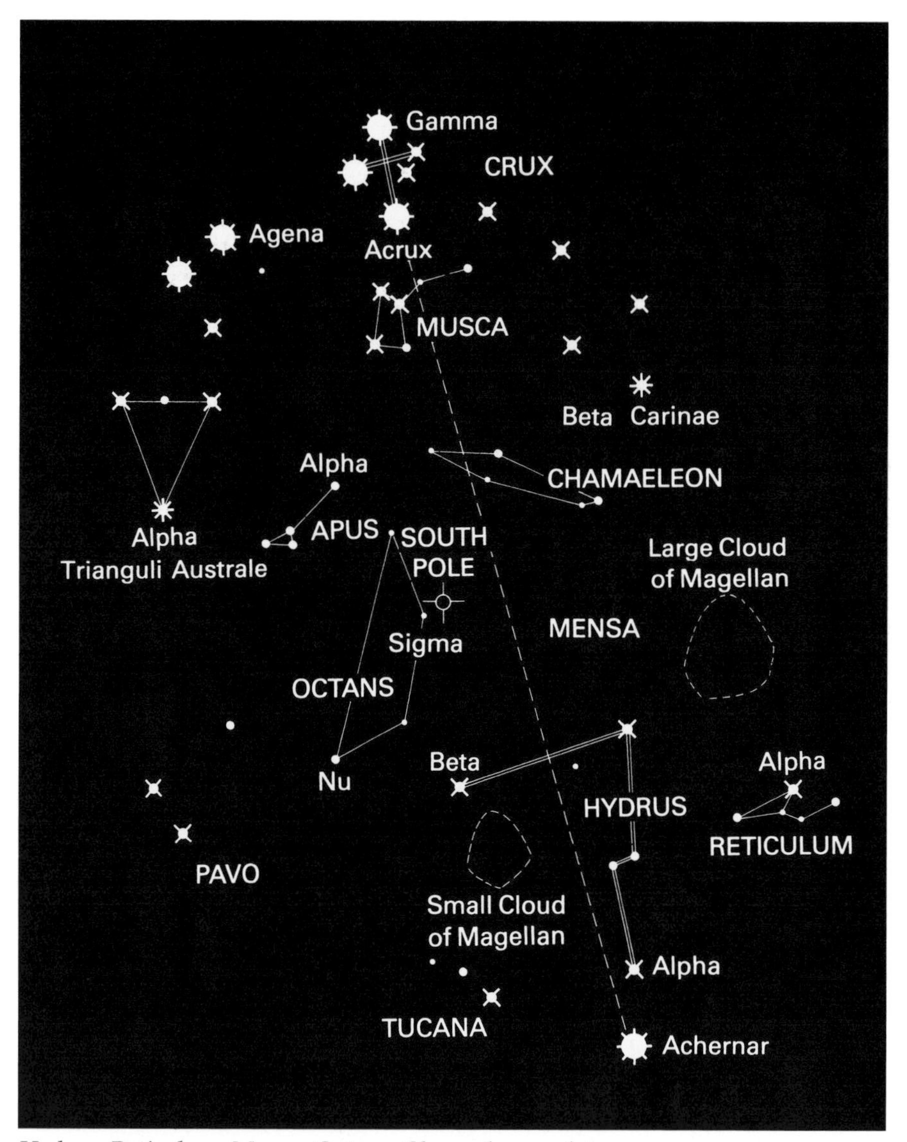

Hydrus, Reticulum, Mensa, Octans, Chamaeleon and Apus

HYDRUS: *the Little Snake*

This is a fairly long constellation, notable only because Beta (2.8) is the nearest reasonably bright star to the south pole – though it is still 13 degrees away. Beta Hydri lies roughly between Achernar and Triangulum Australe (the Southern Triangle), but is best found by its nearness to the Small Cloud of Magellan which does overlap Hydrus, though most of it lies in Tucana. The Little Snake is devoid of interesting naked-eye objects. Alpha (2.9) lies close to Achernar; this is probably the best way to identify Hydrus.

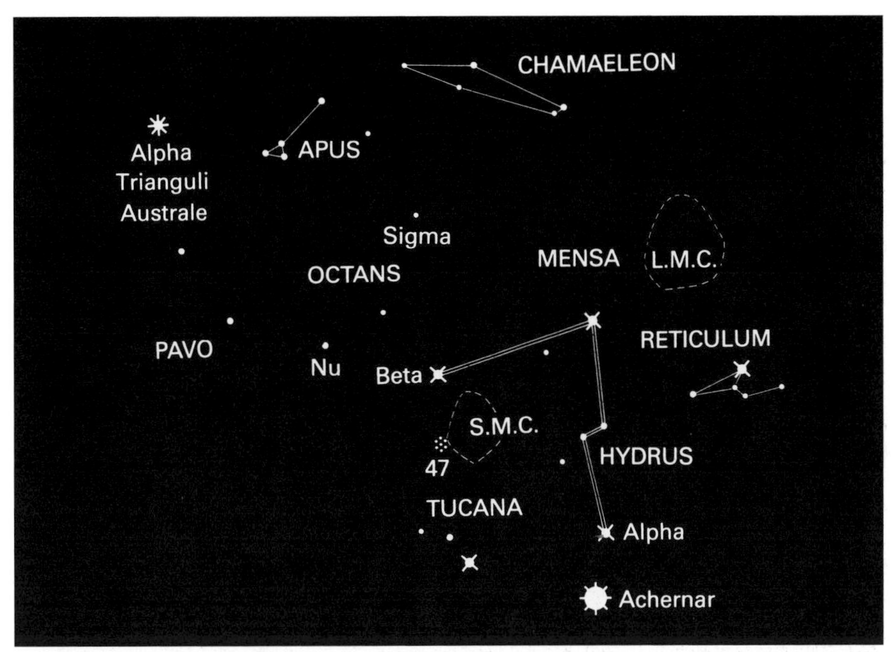

Octans, Apus, Chamaeleon, Mensa, Reticulum and Hydrus

RETICULUM: *the Net*

Reticulum contains no star brighter than Alpha (3.3), but it is quite compact and is not hard to find. Look for it midway between Achernar and Canopus. Like so many of these far southern constellations, it contains nothing of note.

MENSA: *the Table*

Mensa (originally Mons Mensae, the Table Mountain) has the dubious distinction of being the only constellation in the sky with no star as bright as the fifth magnitude. However, it does contain a small part of the Large Cloud of Magellan. There seems no justification for its existence as a separate group.

OCTANS: *the Octant*

Octans, the south polar constellation, is very barren indeed. Its brightest star, Nu, is only of magnitude 3.8.

Sigma Octantis, the actual Pole Star, is of magnitude 5.5, so that it is never a really easy naked-eye object, and is hidden by the slightest haze. City-dwellers have no chance of seeing it at all with the naked eye. Moreover,

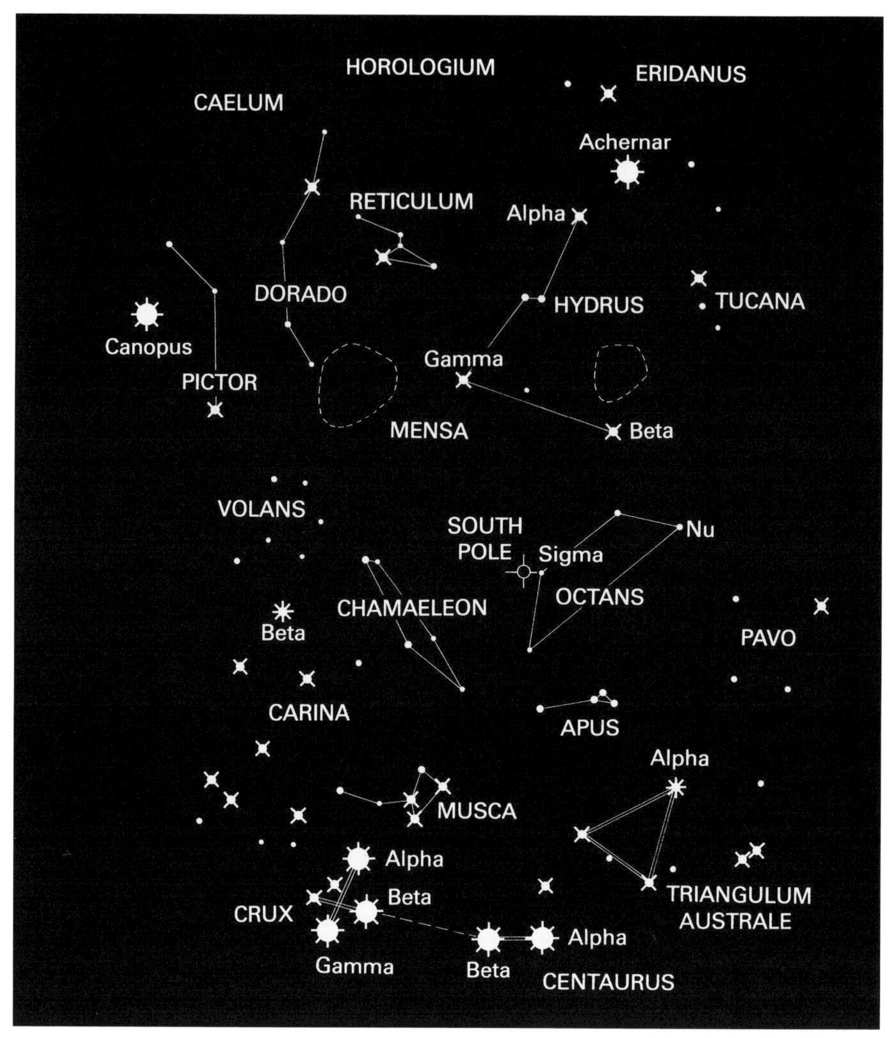

Stars of the far south

there is nothing distinctive anywhere near it, and to locate it I am afraid that you will have to do some careful searching; I know that when I first went to the southern hemisphere I had to examine the area for at least half an hour before I was certain that I had identified Sigma. At the end of the twentieth century it is over a degree of arc from the pole.

CHAMAELEON: *the Chameleon*

This is yet another small, dim group, with no star as bright as the fourth magnitude. It has the shape of a distorted diamond, but is not easy to identify, and it is entirely uninteresting.

APUS: *the Bee*

This was originally Avis Indica, the Bird of Paradise. Its brightest stars are only marginally above the fourth magnitude. To find it, first locate Alpha Trianguli Australe and then look in the direction of the Large Cloud of Magellan. Apus too contains nothing of note.

THE MAGELLANIC CLOUDS

The Magellanic Clouds, or Nubeculae, are among the most important features of the sky. Both lie in the far south, and it seems best to deal with them here, though the Small Cloud lies mainly in Tucana and the Large Cloud in Dorado.

They are much the brightest of the external galaxies; the Large Cloud is 169,000 light-years away, and the Small Cloud about 190,000 light-years. Outwardly they look like broken-off parts of the Milky Way. Both are irregular in shape, though the Large Cloud shows vague indications of barred spiral structure, while the Small Cloud may just possibly be made up of two separate parts in almost the same line of sight. Both are enveloped in a huge mass of very tenuous gas, which may even extend through to the far side of our Galaxy. Both Clouds may be regarded as satellite systems of our Galaxy. The Large Cloud is a quarter the size of our Galaxy, and the Small Cloud one-sixth the size. This means that they are by no means dwarf systems.

Their importance lies in the fact that they contain objects of all kinds – giant and dwarf stars, variables, doubles, open and globular clusters, nebulae, and fairly frequent novae. Both are conspicuous; the Large Cloud remains visible even in moonlight. They were first well described by the explorer Magellan in 1519 during his voyage round the world, but they must certainly have been noticed earlier. The Large Cloud contains the Tarantula Nebula, which is an easy object for the naked eye and is not far short of a thousand light-years in diameter; this means that it is far larger than the Orion Nebula.

In 1987 a supernova flared up in the Large Cloud, and attained the second magnitude, so that it was a prominent naked-eye object and altered the aspect of the whole of that part of the sky. This was the first naked-eye supernova seen since Kepler's Star of 1604. It has now become a very dim telescopic object.

The two Clouds are about 80,000 light-years apart, centre to centre. Only two known galaxies are closer to us; one is a dwarf system in Orion, 70,000 light-years away and much too faint to be seen with the naked eye – it has even been given the somewhat derisive nickname of 'Snickers'! The other is a dwarf system in Sagittarius.

20

Southern stars: the January sky

For our seasonal charts, I have chosen a latitude of 35 degrees south, which means that they apply to much of South America, South Africa, Australia and New Zealand, though there are of course noticeable differences; thus Ursa Major rises from Darwin but never from Wellington. I have drawn the maps for 10 p.m. There seems little point in giving G.M.T. as well, because so many time-zones are involved. The charts are valid for the following dates and times:

1 October: 4 a.m.	1 January: 10 p.m.
1 November: 2 a.m.	1 February: 8 p.m.
1 December: midnight	1 March: 6 p.m.

Orion is fortunately high. Now the Belt points upward to Sirius, downward to Aldebaran; Capella is very low in the north, and a line extended from Mintaka through Bellatrix will lead to Canopus, the brightest of all the stars apart from Sirius, which is quite unmistakable and is high up. The Southern Cross is rather low in the south-east and Fomalhaut dropping in the south-west, with Achernar not very far from the zenith; the Twins are rising in the north-east, and Perseus is visible low in the north-west. The Milky Way is clearly visible, running from Crux through to Perseus.

I have already discussed the constellations visible from northern latitudes, and it seems pointless to repeat the descriptions here – but remember that everything is reversed, so that, for instance, Rigel in Orion is higher than Betelgeux. This is always an awkward problem in any account of the entire sky; I hope that it will not lead to any confusion.

STAR MAP 13

35° S 15 Jan 10 p.m.

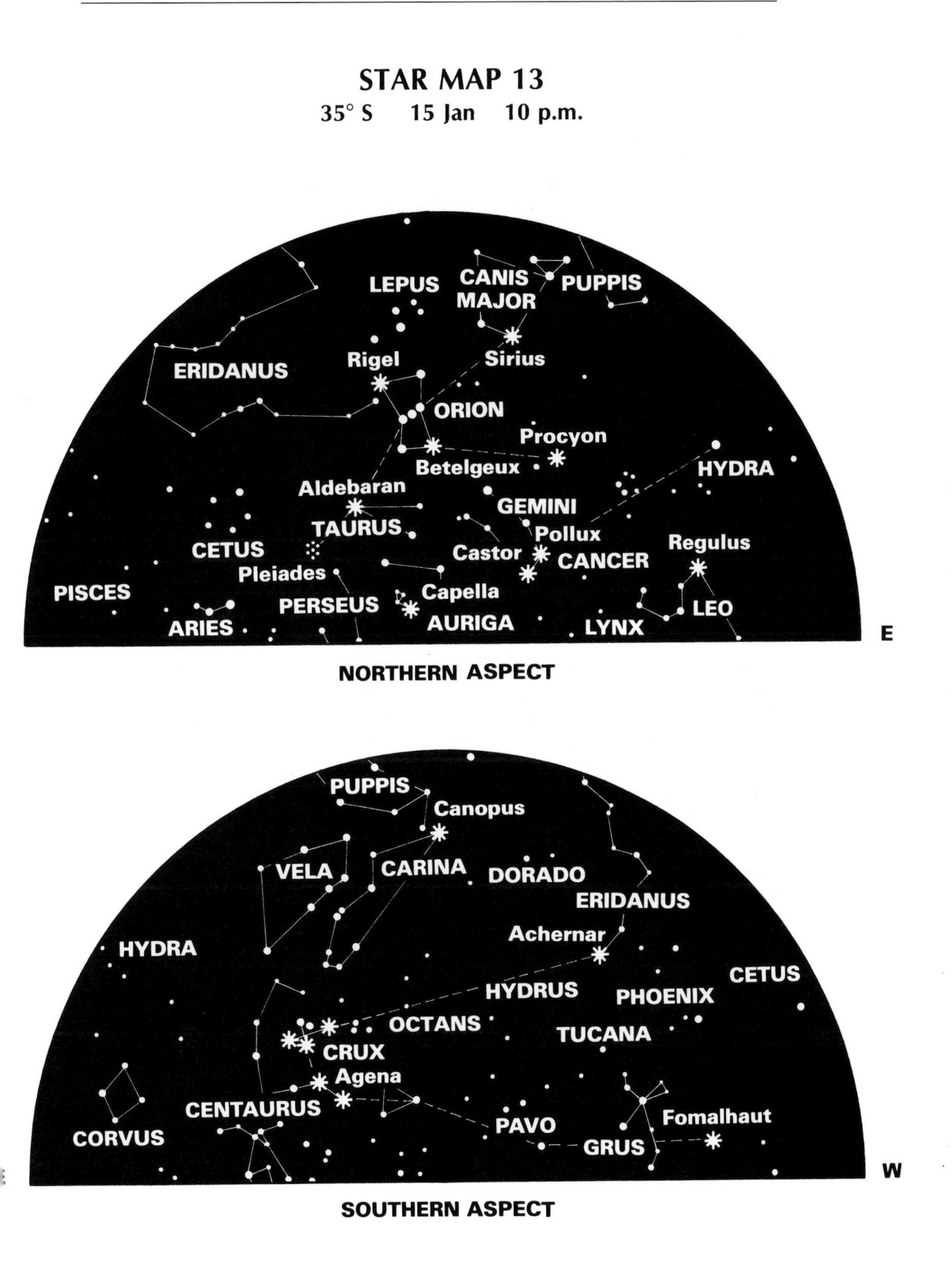

CARINA: *the Keel*

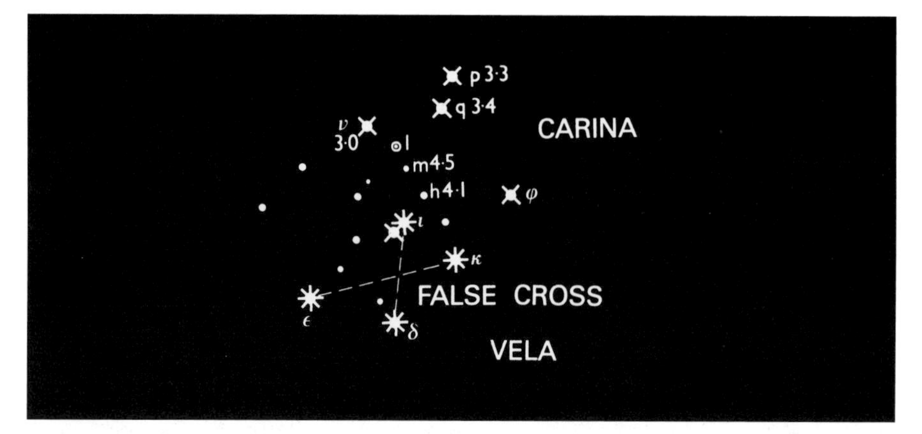

ι Carinae

Carina is the most brilliant part of the now-dismembered Argo Navis. Canopus can be found from Orion, though it is in any case so bright that it can hardly be missed. It is said to be slightly yellowish, but to me it always looks pure white, and I very much doubt whether any colour will be seen in it either with the naked eye or with a telescope. It is very remote – so remote, indeed, that its distance and luminosity are uncertain. It has been said to be 200,000 times as powerful as the Sun. This may be a considerable overestimate, but at any rate Canopus ranks as one of our cosmic searchlights.

The second brightest star in the constellation is Beta (1.7), which lies roughly between Crux to the one side and Canopus to the other. Theta Carinae (2.8) is the central star of a large, loose cluster, IC 2602 (Caldwell 102), which is not very evident with the naked eye, though it can be seen that Theta itself looks rather hazy. (Note that there are no Messier numbers for these southern clusters and nebulae, because Messier lived in France, and could never see them. IC stands for Index Catalogue; NGC for New General Catalogue – though it is no longer new, for it was drawn up by the Danish astronomer J. L. E. Dreyer almost a century ago.) There are two other naked-eye clusters in Carina: NGC 2516 (Caldwell 96), near Epsilon, and NGC 3114, which forms a triangle with Iota Carinae (2.2) and Kappa Velorum in the False Cross. Both are rich, and should be identified without difficulty. Not far from Iota is the Cepheid variable I Carinae, with a range of 3.4 to 4.8 and the unusually long period, for a Cepheid, of 35 days. Beside it is the Mira variable R Carinae, which can reach the fourth magnitude at maximum, though at minimum it fades to 10; it is visible with the naked eye for a week or two each year (its period is 381 days) but I always find it rather difficult to locate.

Iota and Epsilon Carinae (1.9) make up the False Cross, together with Kappa Velorum (2.5) and Delta Velorum (2.0). Three of its stars are white; the other, Epsilon Carinae, is obviously orange.

VELA: *the Sails*

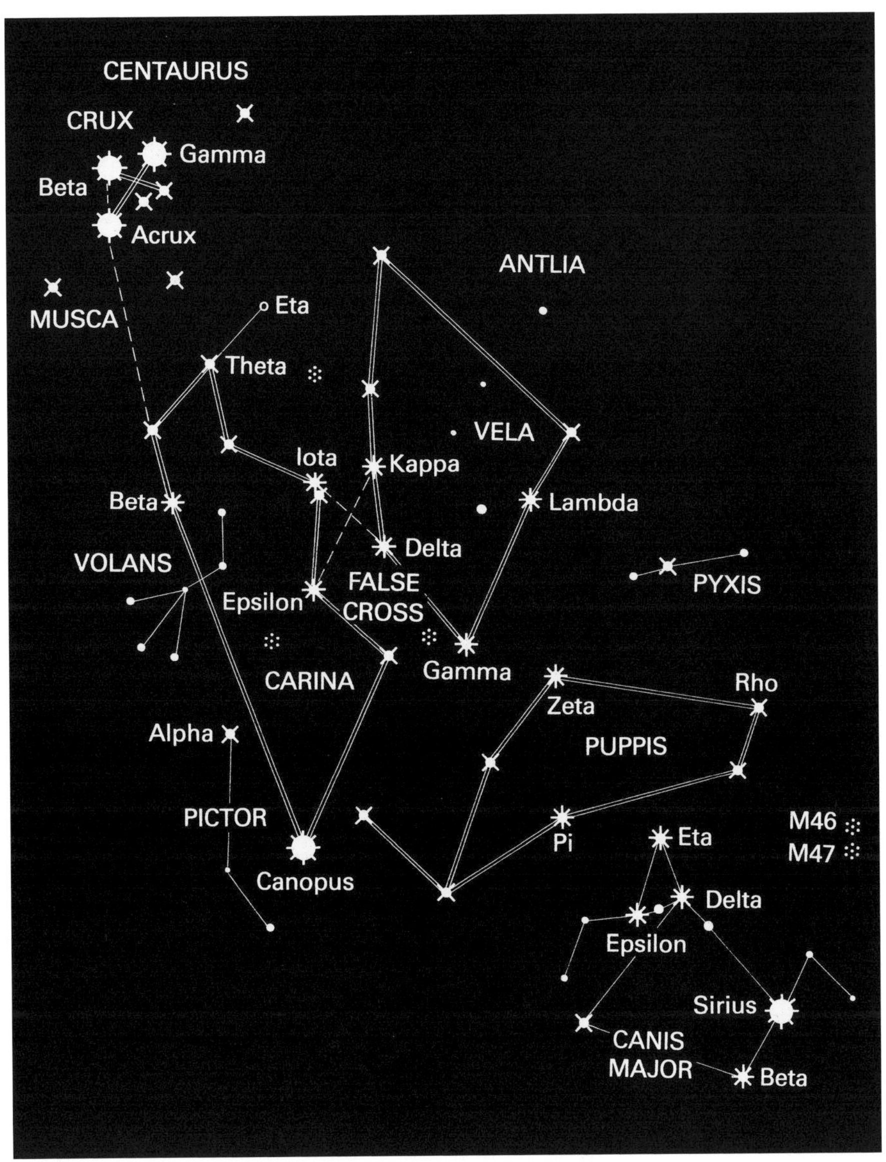

Carina, Vela, Puppis, Pyxis, Volans, Pictor and Antlia

For obvious reasons, I have shown the Sails of the old Argo together with the Keel. It is worth noting that Delta Velorum and Iota Carinae point one way to the Theta Carinae cluster, and the other way to Gamma Velorum (1.8), which is an exceptionally hot star of unusual type. The other bright star of Vela, Lambda (2.2), forms a triangle with Delta and Gamma. The Milky Way flows through the constellation, and the open cluster NGC 2547, close to Gamma, is just visible with the naked eye.

PUPPIS: *the Poop*

This is the only part of Argo ever visible from British or Canadian latitudes, and the brightest star, Zeta (2.2), does not rise, though it can be seen from New York very low over the horizon. It is an extremely hot and powerful star, 60,000 times as luminous as the Sun. If it were as close to us as Vega, it would appear twelve times as brilliant as Venus!

There are two naked-eye variables in Puppis; L^2, close to Sigma (3.3), has a range of 3.4 to 6.2, so that at minimum it becomes too faint to be seen without optical aid, and the Beta Lyrae type eclipsing binary V, close to Gamma Velorum and only just inside the boundary of the Poop. V Puppis has a period of 1.4 days, and varies between magnitudes 4.3 and 5.1.

The open cluster M. 47, in the northernmost part of Puppis, is easy to find. Actually the best way to locate it is by using Sirius and Gamma Canis Majoris as pointers. Close by it is another open cluster, M. 46, which is said to be visible with the naked eye though I have never been able to detect it without binoculars.

PYXIS: *the Compass*

This is the last section of Argo. It has no star above magnitude 3.5 and is of no interest, but it can be found between Rho Puppis (2.8) and Lambda Velorum.

VOLANS: *the Flying Fish*

Volans seems to be an unnecessary constellation; it intrudes into Carina, between Beta and Canopus. Its four brightest stars are between magnitudes 3.5 and 4.

PICTOR: *the Painter*

An obscure group with no bright stars. Beta Pictoris (3.8) is interesting because it has been found to be associated with cool material, and may well be the centre of a system of planets, but outwardly it looks normal enough. It is 78 light-years away and 60 times as luminous as the Sun. Pictor's three main stars, all below the third magnitude, lie in a line close to Canopus.

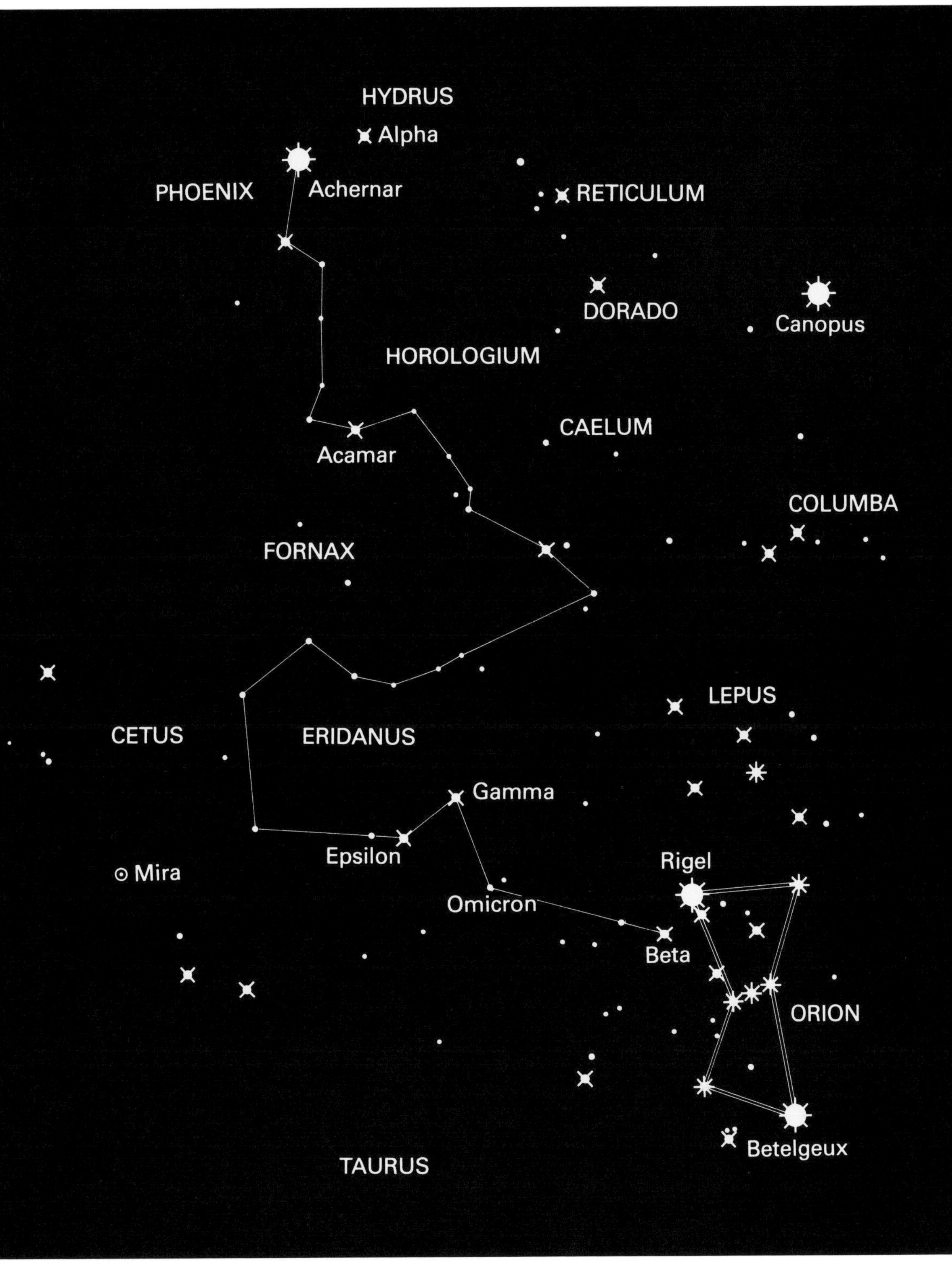

Eridanus, Dorado, Caelum and Horologium

ERIDANUS: *the River*

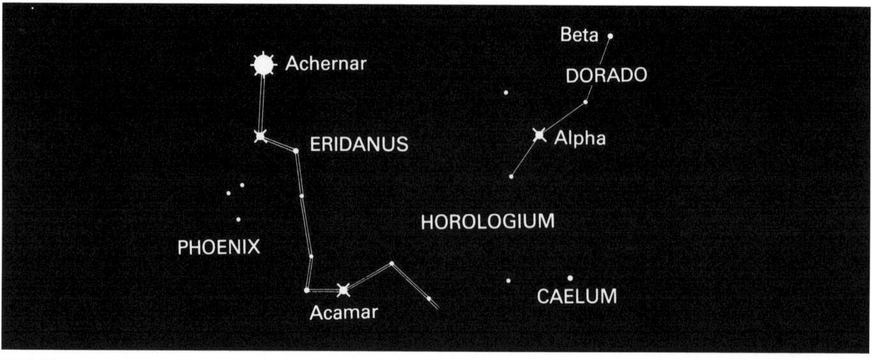

Eridanus

Eridanus has already been described in part, but its most brilliant star, Achernar (0.5), is not too far from the south pole. To see it all you must go south of latitude 30 degrees north – so that it rises from Delhi, but not from Athens or San Francisco. It is a white star, 85 light-years away and well over 700 times as luminous as the Sun. From it you can easily trace the line of stars marking the River, which during January evenings stretches up almost overhead, ending near Rigel. Theta Eridani or Acamar (2.9) was ranked as of the first magnitude by the ancient astronomers, though whether it has actually faded is doubtful. It is a fine binary, separable with a small telescope but not with the naked eye.

HOROLOGIUM: *the Clock*

Horologium is a very obscure group, with no star brighter than magnitude 3.9. It lies roughly between Achernar and Lepus, so that I have marked its position on the map with Eridanus.

DORADO: *the Swordfish*

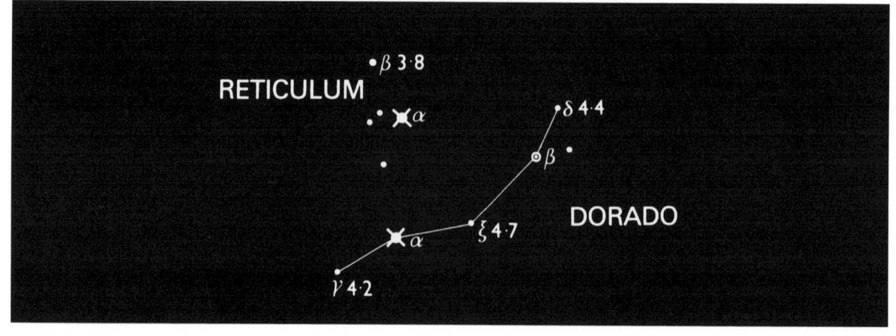

β Doradûs

Dorado, lying between Achernar and Canopus, has only one non-variable star, Alpha (3.3), above the fourth magnitude; it adjoins Reticulum. Beta is a Cepheid variable, with a period of 9.8 days and a magnitude range of 3.8 to 4.7, so that it is always visible with the naked eye; a good comparison star is Delta (4.4). Dorado contains part of the Large Cloud of Magellan, including the magnificent Tarantula Nebula, known officially as 30 Doradûs or NGC 2070 (Caldwell 103).

21

Southern stars: the February sky

The chart for this month is valid for;

1 November: 4 a.m
1 December: 2 a.m.
1 January: midnight
1 February: 10 p.m.
1 March: 8 p.m.

Canopus is now almost at the zenith, and the main glory of the southern sky, Crux, is gaining altitude in the south-east; it and Achernar, in the south-west, are about equally high, so that this is a good time for locating the south celestial pole. Orion is still excellently placed, and so is Sirius, so that it is interesting to compare Sirius with Canopus; the difference between them is less than a magnitude. Capella remains on view, low in the north though from Invercargill, in the southernmost part of New Zealand, or southern Argentina and Chile, it will not be seen. Fomalhaut is setting in the south-west and Leo rising in the north-east; Procyon and the Twins are prominent, and the Milky Way is splendidly displayed. Cetus lies in the west. Since we have no Pegasus to help in locating it, the best method is to extend the line from Crux through Achernar almost down to the western horizon. Alphard, the 'Solitary One' in Hydra, can be identified because it lies inside the large triangle formed by Sirius, Canopus and Regulus.

STAR MAP 14

35° S 15 Feb 10 p.m.

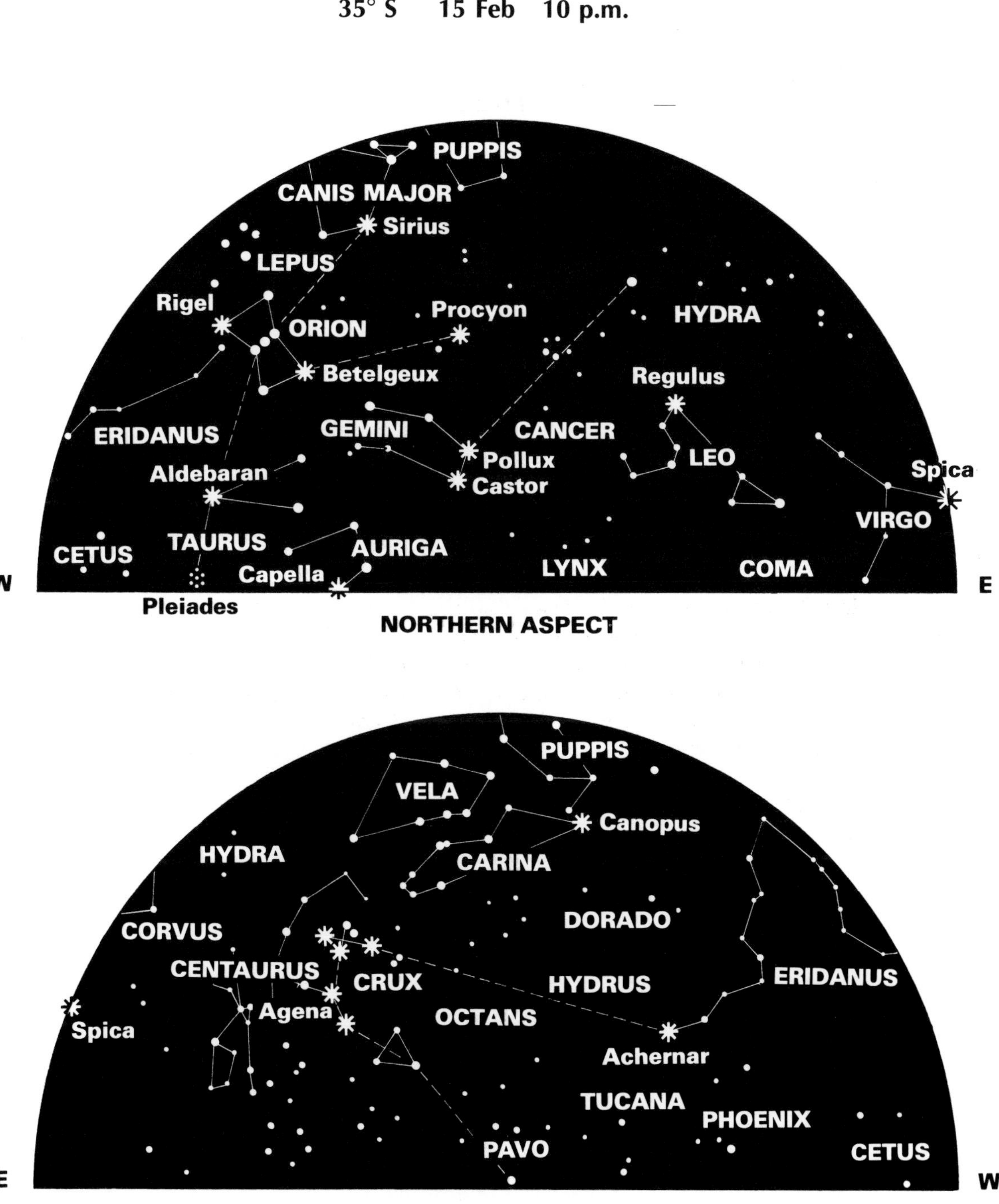

22

Southern stars: the March sky

The two-hour shift makes a noticeable difference to the evening sky. The charts apply to:

1 December:	4 a.m.
1 January:	2 a.m.
1 February:	midnight
1 March:	10 pm.
1 April:	8 p.m.
1 May:	6 p.m.

By now Orion has started to descend in the west, but it is still well above the horizon and remains available as a guide. Sirius and Canopus are very high, and the Southern Cross is now at a greater altitude than Achernar. The Clouds of Magellan are excellently placed, and the Scorpion is starting to come into view – later in the year it will dominate the scene in a way that it can never do from Europe or the northern United States. The Twins lie at a respectable altitude in the north, and Leo is now well clear of horizon mists, with Spica rising in the east. Look for Praesepe, between Regulus and the Twins. The area low in the north is very barren, since it is occupied mainly by Lynx; Capella has set, and Aldebaran is now nearing the horizon. As in February, the Milky Way is very much in evidence.

STAR MAP 15

35° S 15 Mar 10 p.m.

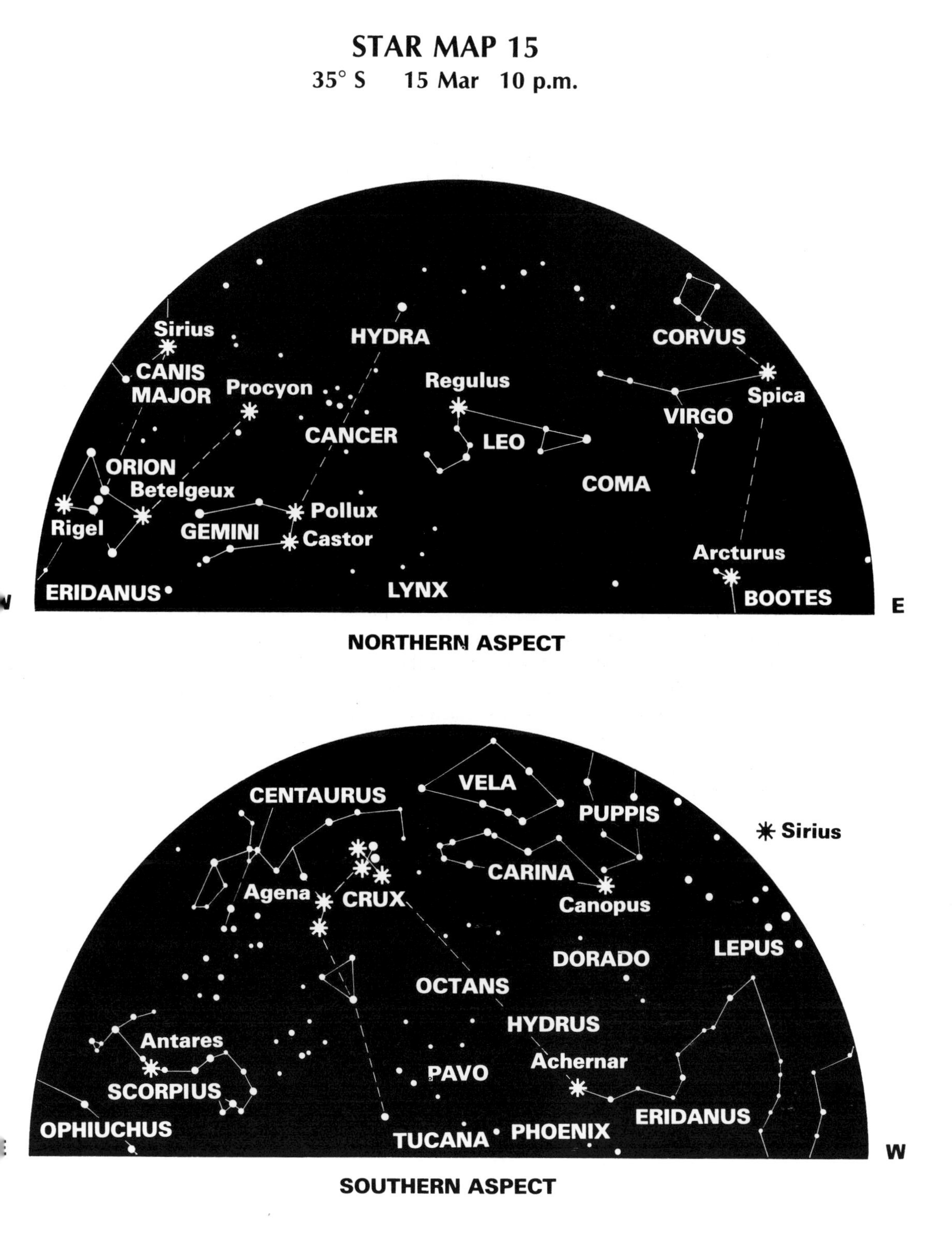

23

Southern stars: the April sky

By evenings in April we have more or less lost our main guide; Orion is setting in the west, and of his retinue only Sirius remains well placed. The charts here are valid for:

1 January:	4 a.m.
1 February:	2 a.m.
1 March:	midnight
1 April:	10 p.m.
1 May:	8 p.m.
1 June:	6 p.m.

The Southern Cross is now almost overhead, together with the Centaur; the Cross and the two Pointers, Alpha and Beta Centauri, are quite unmistakable. Achernar is so low that it may well be lost in horizon mist. Scorpius is rising in the south-east, and its red leader, Antares, is already conspicuous. Leo is high in the north, followed round by Virgo; the barren Hydra occupies much of the sky near the zenith. This is a good time for trying to see the stars of Ursa Major, which is skirting the northern horizon. Arcturus is coming into view in the north-east.

CRUX AUSTRALIS: *the Southern Cross*

The four chief stars are Alpha or Acrux (0.8), Beta (1.2), Gamma (1.6) and Delta (2.8). Beta Crucis, though ranked as a first-magnitude star, has no official proper name; some people call it Mimosa.

Crux is not an ancient constellation. Until 1679 it was included in Centaurus, which almost surrounds it. It was created by an astronomer named Royer, who made no other notable contributions to science.

Of the stars making up the Cross, three are pure white, while the fourth – Gamma – is of a beautiful orange colour, as even a casual glance will show. Binoculars bring out the contrast even more clearly.

The magnificent cluster Kappa Crucis (C. 94), close to Beta, is known as the Jewel Box, because with a telescope it contains many blue-white stars and one prominent red supergiant. With the naked eye, it appears as a misty patch. Immediately to the south of it lies the Coal Sack (C. 99), an apparently

STAR MAP 16

35° S 15 Apr 10 p.m.

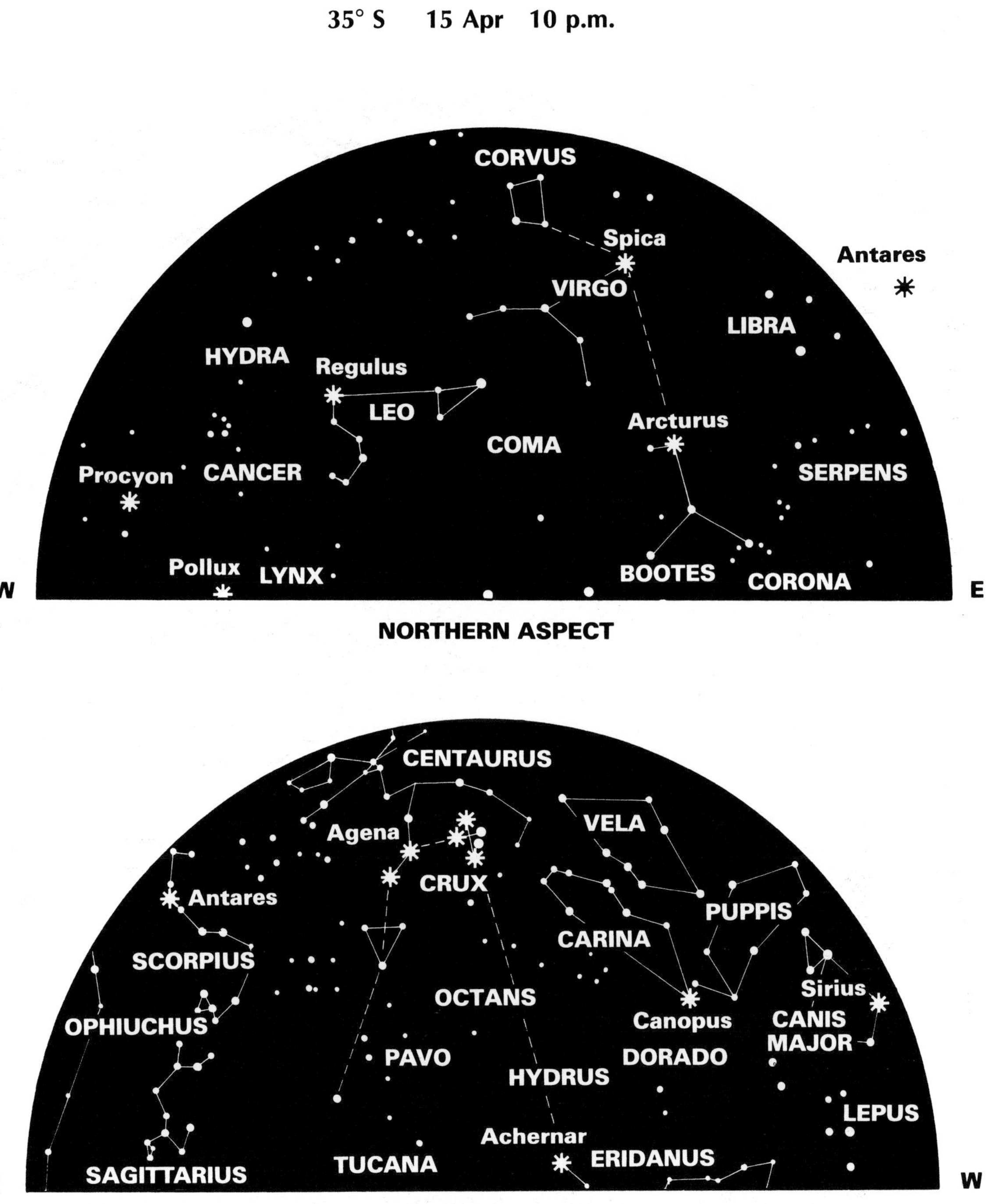

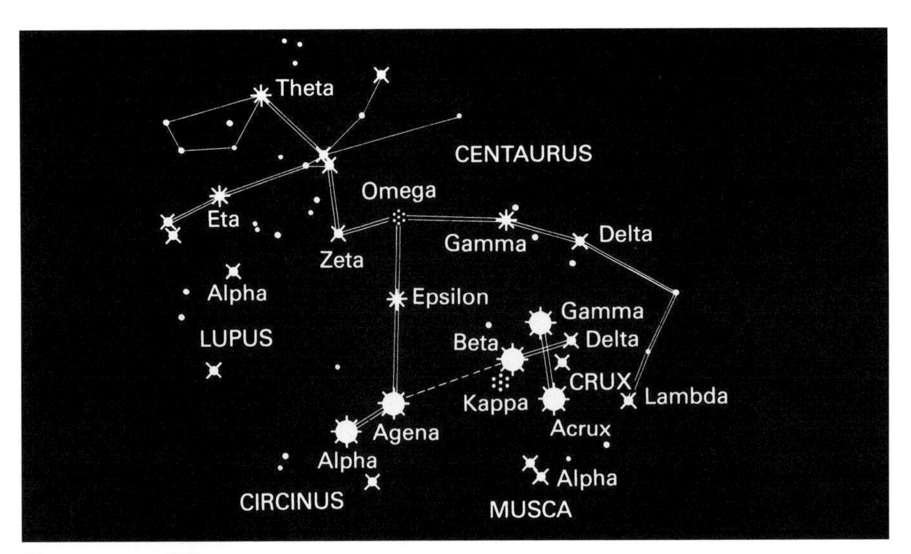

Centaurus and Crux

starless area which is in fact due to dark nebulosity – unlit by any star as seen from our vantage point in the Galaxy, though for all we know its far side may be illuminated. It measures 7 by 5 degrees and is easy to find, near Beta and just east of Acrux. It seems to be about 70 light-years in diameter and no more than 600 light-years away, so that there are not many foreground stars to be seen even with large instruments.

CENTAURUS: *the Centaur*

Centaurus is one of the most splendid of all the constellations. Alpha and Beta point straight to Crux, and are known as the Southern Pointers. They are not alike; Alpha, magnitude −0.3, is the brightest star in the sky apart from Sirius and Canopus, but is only just over 4 light-years away; its dim companion, Proxima, is actually the closest of all the stars beyond the Sun. Beta, known by its proper name of Agena, is 460 light-years from us, and is the equal of well over 10,000 Suns, so that the two Pointers have nothing whatsoever to do with each other. Other bright stars in Centaurus are Gamma (2.2), Epsilon (2.3), Eta (2.3) and Zeta (2.5). Theta (2.1) is orange, and is bright enough for the colour to be noticeable with the naked eye. It is sufficiently far north to rise over parts of Europe and Canada.

Centaurus includes much the finest globular cluster in the sky, Omega (C. 80), which is prominent with the naked eye as a fuzzy blur. To find it, simply follow up the line from Beta through Epsilon, and continue it for an equal distance on the far side. Its distance is around 17,000 light-years; its real diameter is about 150 light-years, and probably it contains at least a million stars. Near its centre, the average separation between the stars is no more than a tenth of a light-year; even so, direct collisions must be very rare indeed.

LUPUS: *the Wolf*

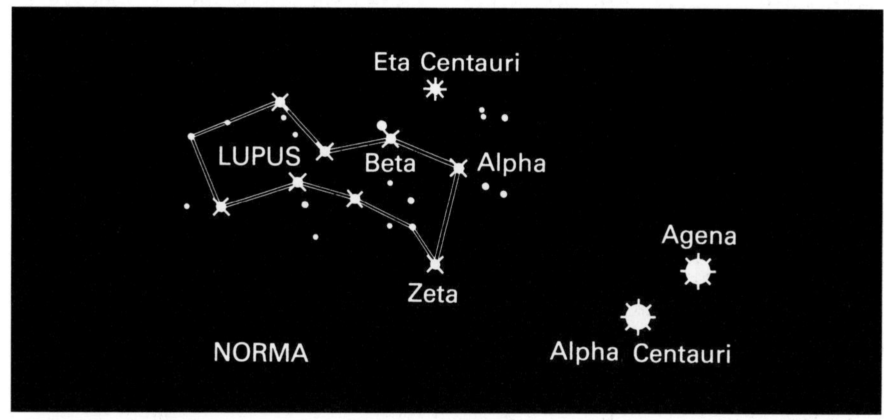

Lupus and Norma

Theta and Eta Centauri act as pointers to Lupus, which has three stars above the third magnitude (Alpha, 2.3, is the brightest) but has no obvious shape, so that you will have to spend some time in sorting it out. It contains no naked-eye objects of note.

MUSCA AUSTRALIS: *the Southern Fly*

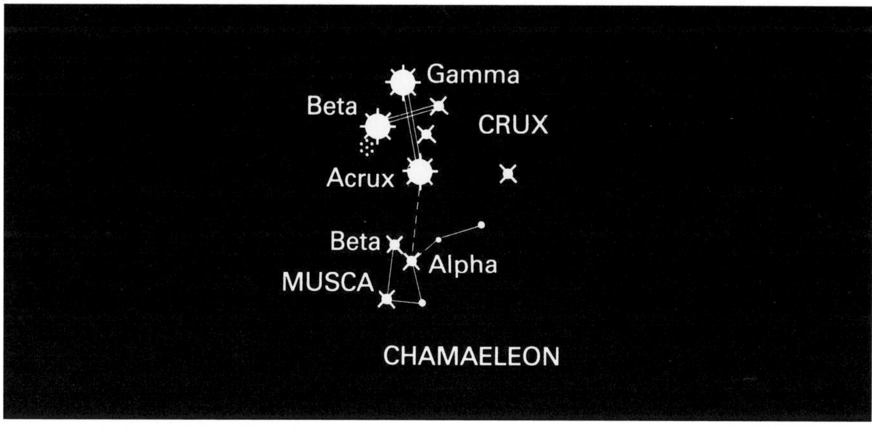

Crux and Musca Australis

Musca is a small but fairly compact constellation; it may be found by extending the line from Gamma Crucis through Acrux, and curving it slightly. It has one star above the third magnitude (Alpha, 2.7) but little of interest to the naked-eye observer.

NORMA: *the Rule*

An obscure and frankly unnecessary constellation, adjoining Lupus, Norma contains no objects of interest, and no star above the fourth magnitude, so that it need detain us no further.

CIRCINUS: *the Compasses*

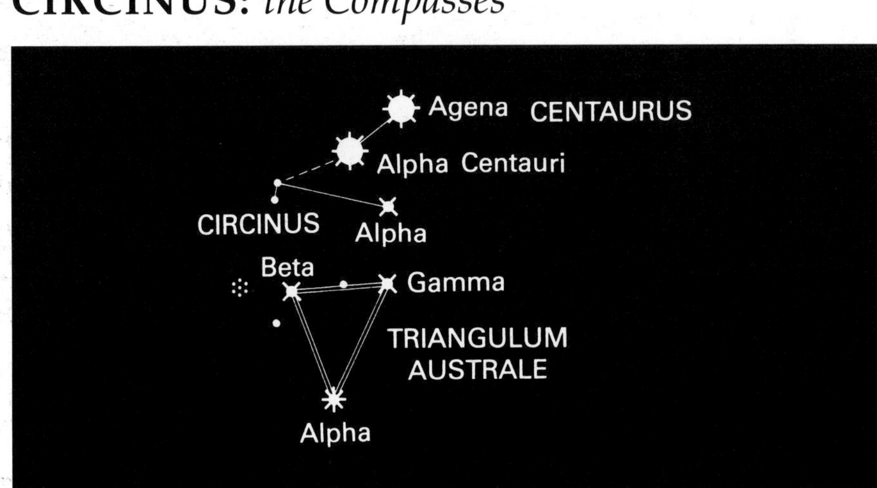

Circinus and Triangulum Australe

Yet another small and unimportant constellation; its only fairly bright star is Alpha (3.2). To find Circinus, extend a line from Beta Centauri through Alpha, and continue it for a short distance on the far side.

TRIANGULUM AUSTRALE: *the Southern Triangle*

Here we have a recognizable constellation. The chief stars, Alpha (1.9), Beta (2.8) and Gamma (2.9), do indeed form a triangle. To find Alpha, an obviously orange-red star, extend a line from Epsilon Centauri through Alpha Centauri. The open cluster NGC 6025 (C. 95) lies in line with Gamma and Beta; it is said to be visible with the naked eye, though I admit that I have never been able to see it clearly without optical aid.

ARA: *the Altar*

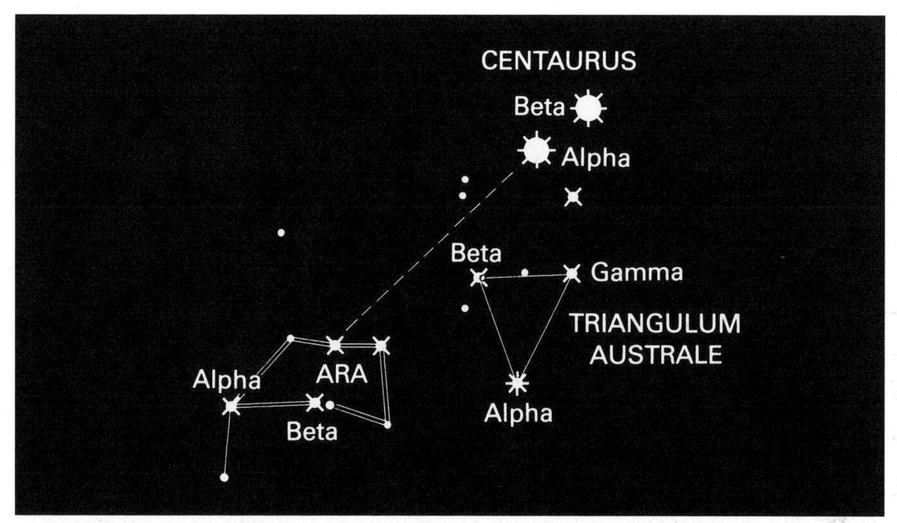

Finding Ara,using Alpha and Beta Centauri and Triangulum Australe

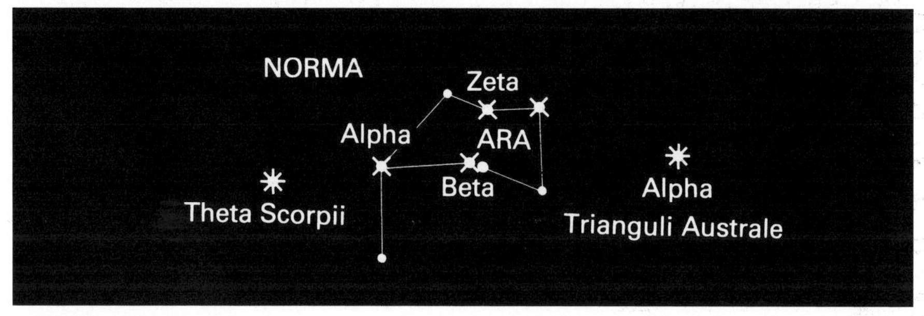

Ara and Norma

Ara may be found by using the Pointers in the direction away from the Southern Cross; note also that Ara lies between Theta Scorpii and Alpha Trianguli Australe. The brightest stars are Beta (2.8), Alpha (2.9) and Zeta (3.1). Zeta is orange, and with the naked eye looks slightly off-white. Ara has a reasonably distinctive shape, but no objects of naked-eye interest.

TELESCOPIUM: *the Telescope*

Telescopium, adjoining Ara, is one more of these small, dim and dull modern constellations. The only star above the fourth magnitude is Alpha (3.5).

24

Southern stars: the May sky

During May evenings the Southern Cross is overhead, and Achernar so low that it will probably not be seen. The maps are valid for:

1 February:	4 a.m.
1 March:	2 a.m.
1 April:	midnight
1 May:	10 p.m.
1 June:	8 p.m.

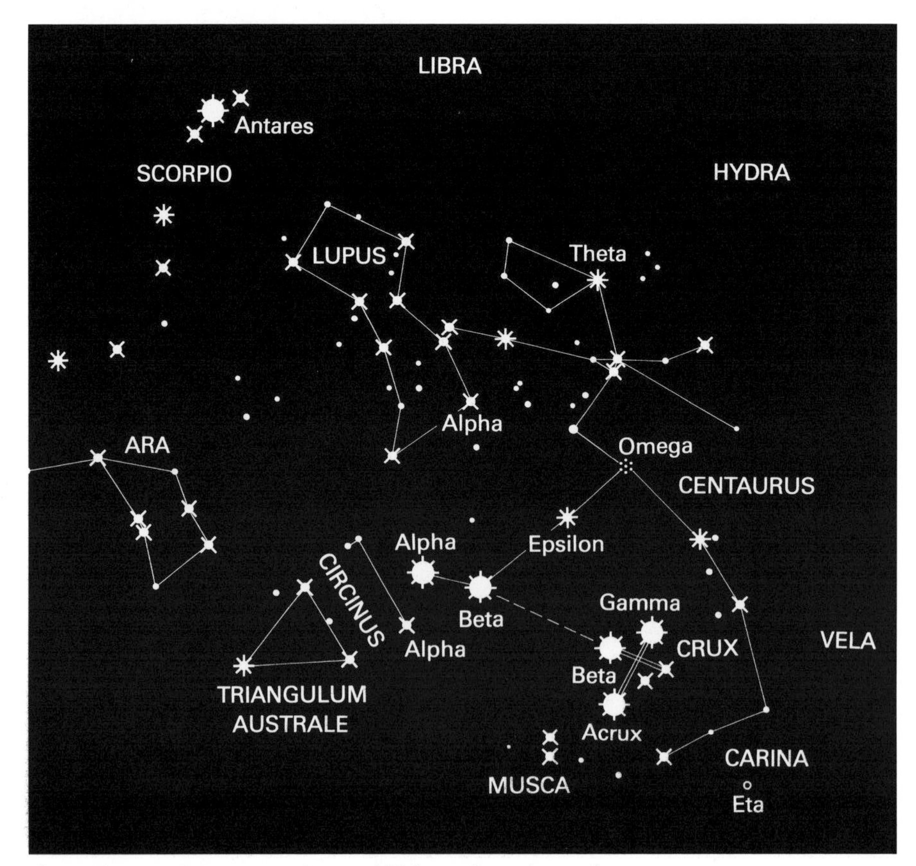

Crux, Lupus, Centaurus, Ara and Triangulum Australe

STAR MAP 17

35° S 15 May 10 p.m.

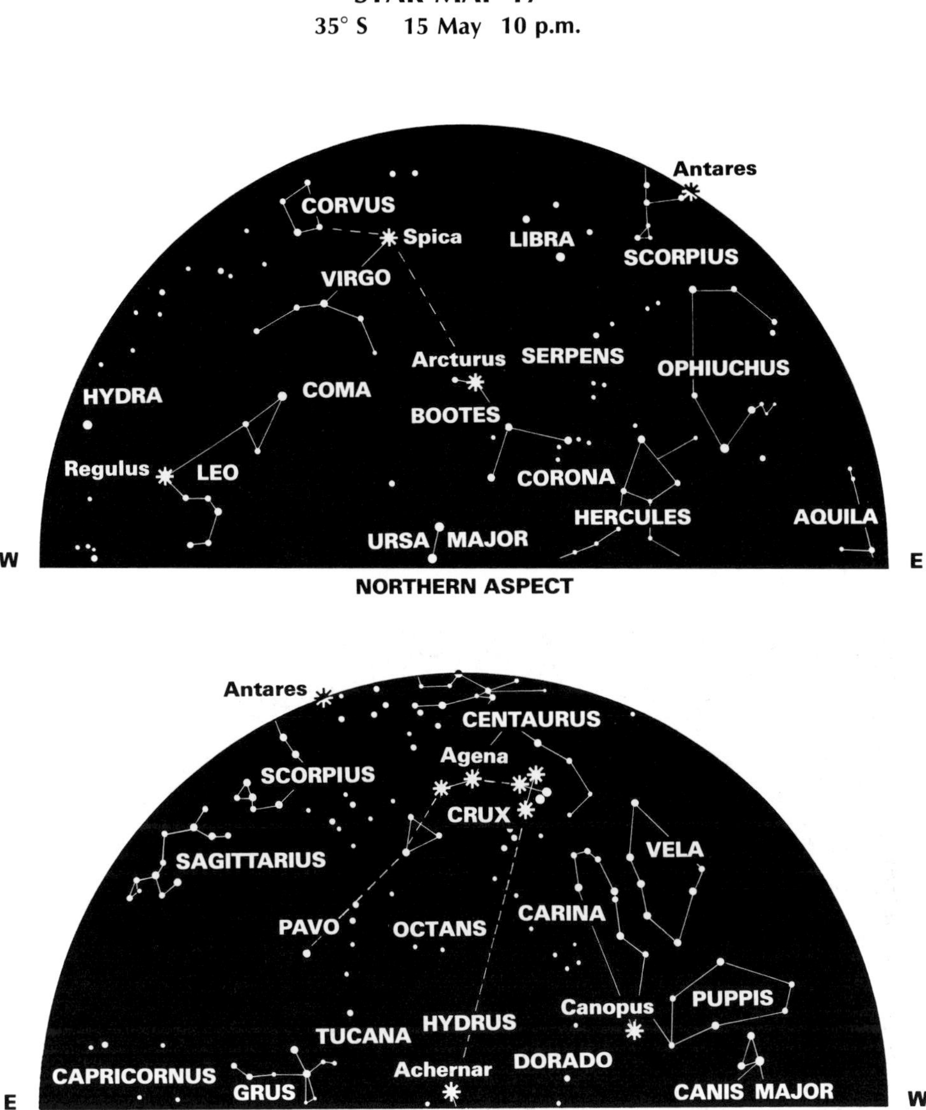

Sirius has now become very low in the south-west, though Canopus is still at a respectable altitude. Scorpius has become prominent in the south-east, and its real glory can now be appreciated. In the north, part of Ursa Major skirts the horizon; Leo and Virgo occupy a large area, and Hydra is higher up, so that this region of the sky appears very barren. Arcturus is well above the north-eastern horizon, and you should be able to make out the semicircle of stars marking Corona Borealis.

The Milky Way is very much in evidence, passing from the south-eastern horizon through Scorpius, Centaurus and Crux, through parts of the old Argo, and then flowing past Sirius down to the south-western horizon.

25

Southern stars: the June sky

The night skies during midwinter contain some brilliant groups, led by the Scorpion, now very high up. The maps apply to:

1 March: 4 a.m.
1 April: 2 a.m.
1 May: midnight
1 June: 10 p.m.
1 July: 8 p.m.

Scorpius is rivalled in splendour by Centaurus, and of course by the Southern Cross, which is slightly south-west of the zenith. Following Scorpius round is Sagittarius, and it is here that we find the very richest part of the Milky Way, with the star-clouds hiding our view of the centre of the Galaxy. On the other hand the northern aspect contains some barren areas – Ophiuchus, Hercules and Serpens; the only bright stars are Arcturus, now at its best, and Spica. Ursa Major has vanished, and Leo is setting in the west. The Clouds of Magellan are at their lowest in the south, though of course they never set over a good part of South America, or over South Africa, Australia or New Zealand.

STAR MAP 18

35° S 15 Jun 10 p.m.

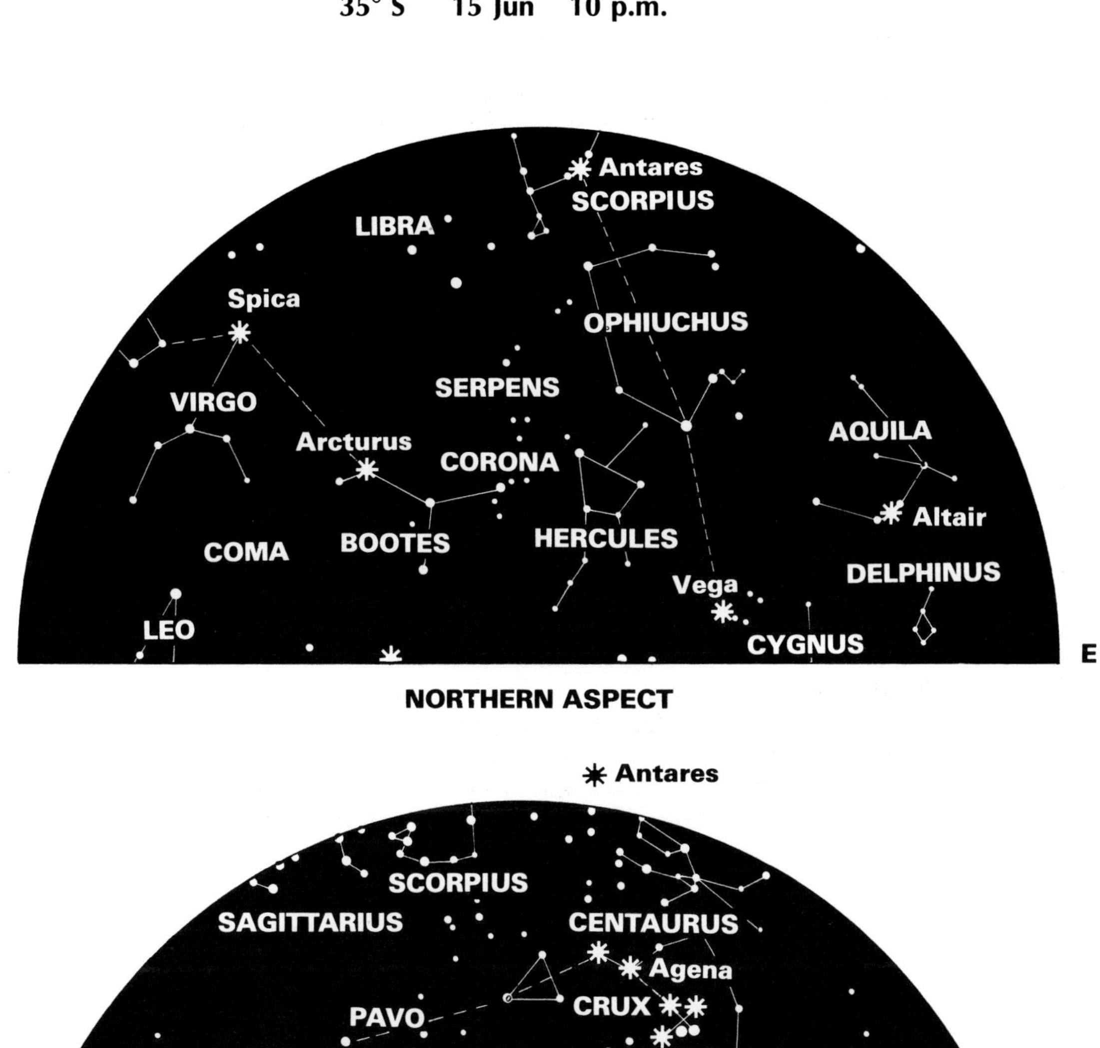

NORTHERN ASPECT

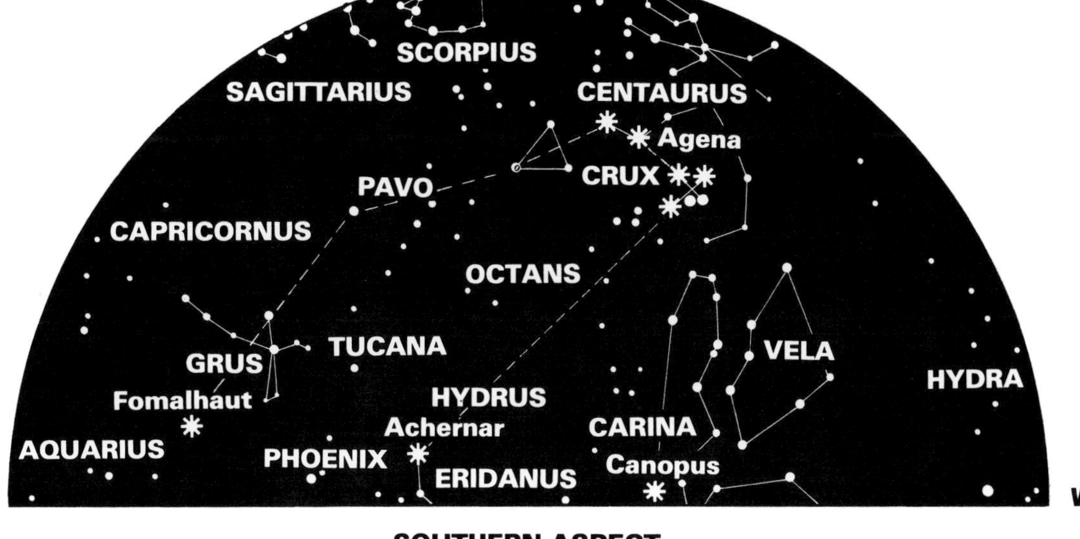

SOUTHERN ASPECT

26

Southern stars: the July sky

The scene is now dominated by Scorpius, which is overhead; the long line of bright stars, led by Antares, cannot possibly be overlooked. The maps have been drawn for:

1 April: 4 a.m.
1 May: 2 a.m.
1 June: midnight
1 July: 10 p.m.
1 August: 8 p.m.

Sirius has now disappeared, and so to all intents and purposes has Canopus. Arcturus is still on view in the north-west, and Corona Borealis can easily be found; Spica is still at a reasonable altitude, but Leo has gone altogether, and once more much of the northern aspect is filled by the barren constellations of the Hercules area. To compensate for this, Vega is now visible, rather low in the north-east, and so is Altair; these are of course two of the stars of the so-called Summer Triangle which so far as southern hemisphere observers are concerned is a Winter Triangle, though the third member, Deneb, has yet to appear. Fomalhaut is rising in the south-east. The Milky Way is now at its very best, since the richest part of it, in Sagittarius, is almost at the zenith; the star-clouds there are impressive even with the naked eye.

All the July constellations have already been described, but it is interesting to show Scorpius and Sagittarius as they are seen from southern latitudes. The Scorpion's Sting is very prominent indeed; Lambda or Shaula is only just below the first magnitude, and makes a notable pair with Upsilon or Lesath (2.7), though the two are not genuinely associated; Lesath is much the more remote and more luminous of the two.

There are two lovely open clusters in Scorpius which are never seen to advantage from the north. Messier 7 is very conspicuous with the naked eye; it lies between Gamma Sagittarii and Shaula – and, incidentally, about mid-way between Antares and the Southern Crown. M. 6, not far away from M. 7, is less brilliant, but is still an easy naked-eye object; with a telescope, the shape has been said to bear some resemblance to a butterfly, but without optical aid it appears simply as a faint blur against a dark sky.

STAR MAP 19

35° S 15 Jul 10 p.m.

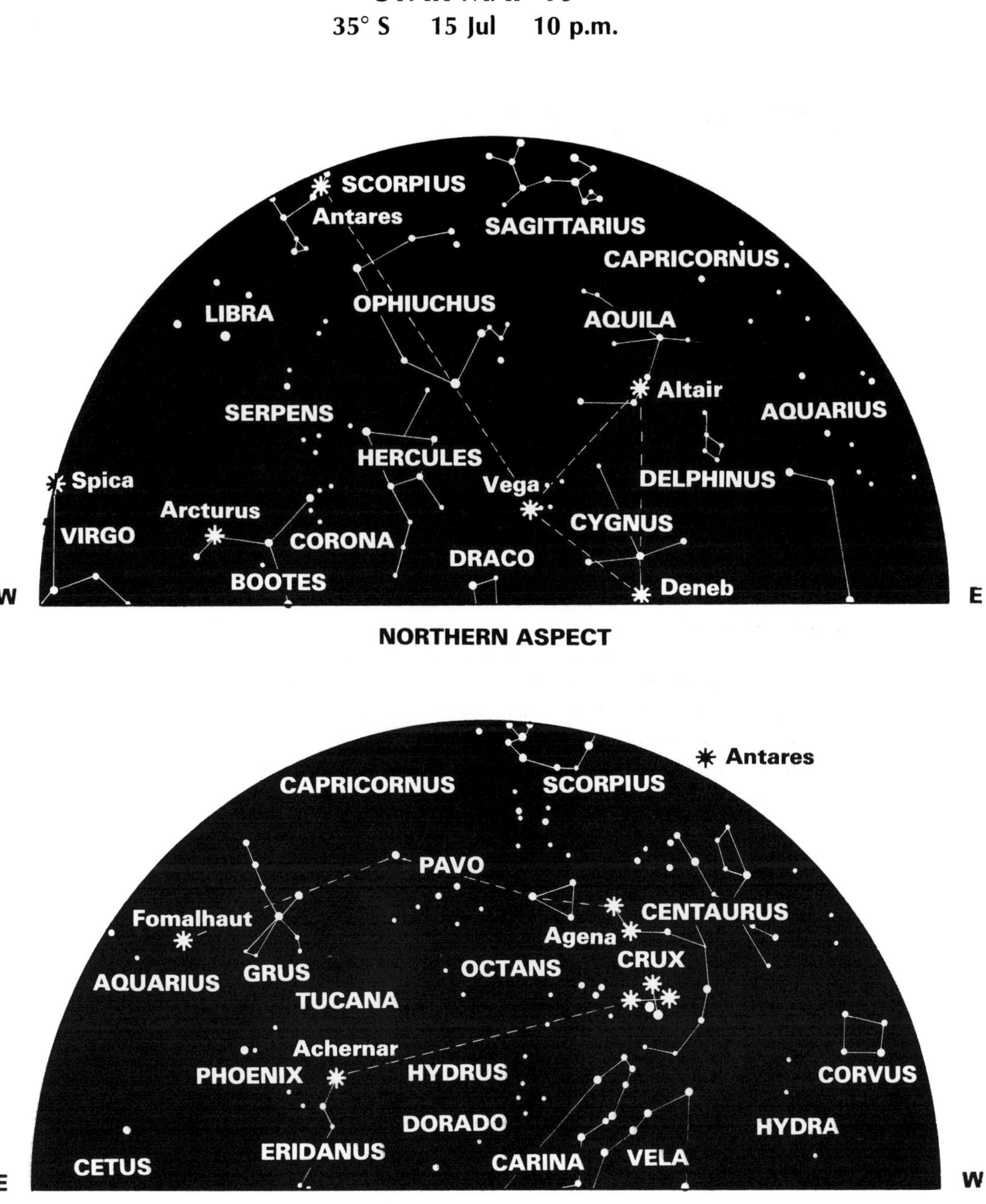

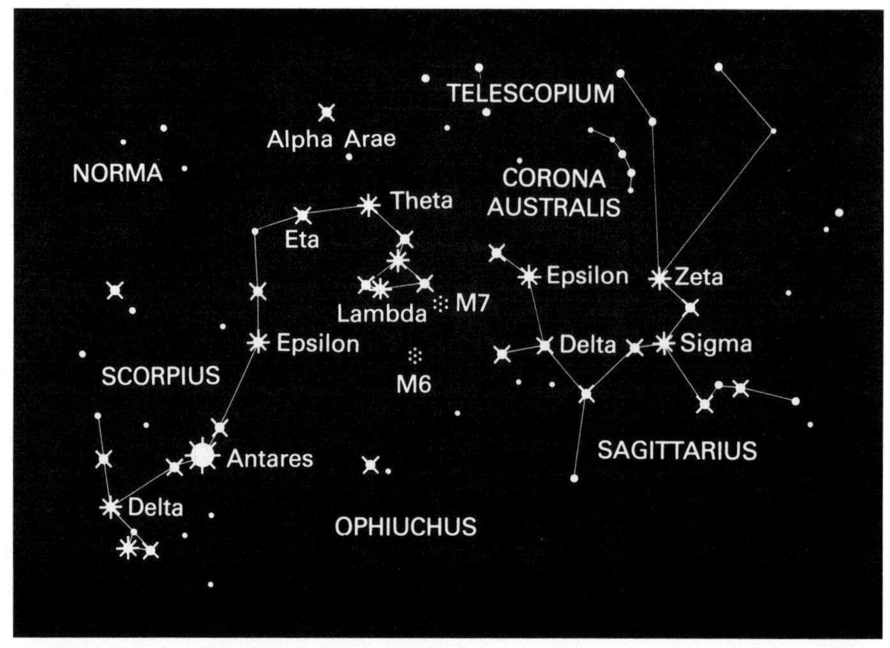

Scorpius, Corona Australis and Sagittarius

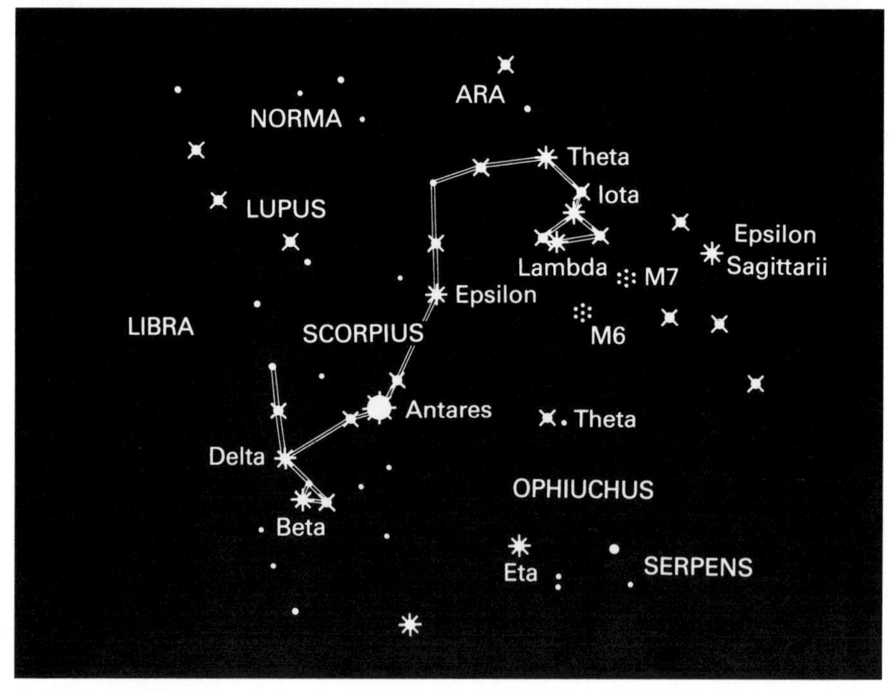

Scorpius and Norma

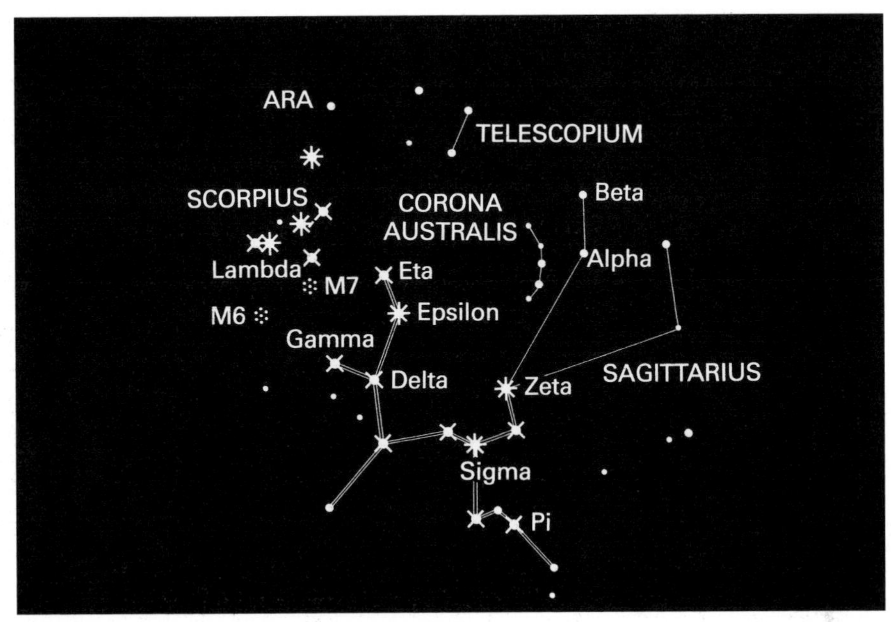

Corona Australis, Telescopium and Sagittarius

27

Southern stars: the August sky

Evening skies in October are still dominated by the Scorpion, which remains very high up, while Sagittarius is practically at the zenith. The maps apply to:

1 May: 4 a.m.
1 June: 2 a.m.
1 July: midnight
1 August: 10 p.m.
1 September: 8 p.m.

Achernar is now well above the south-eastern horizon, and the Southern Cross is dropping in the south-west, so that once more it is easier to locate the decidedly elusive south celestial pole. Spica is low and Arcturus virtually out of view, but in the south-east you will see the rather isolated Fomalhaut: observers who have come from the northern hemisphere are usually surprised to see how bright Fomalhaut is when seen to advantage. Vega is due north, low down; Altair is reasonably high, and the third member of the 'Triangle', Deneb, has risen, though it is always inconveniently low. This is the right time to look for the southernmost stars of Draco. The Dragon's Head may be glimpsed, north-west of Vega, but not at all easily.

As in winter, the Milky Way is superb; the rich region in Cygnus is coming into view, and the shining band may be followed past Aquila, Sagittarius, Scorpius, Centaurus and Crux down to the southern horizon. Canopus is still to all intents and purposes absent from the evening sky.

STAR MAP 20
35° S 15 Aug 10 p.m.

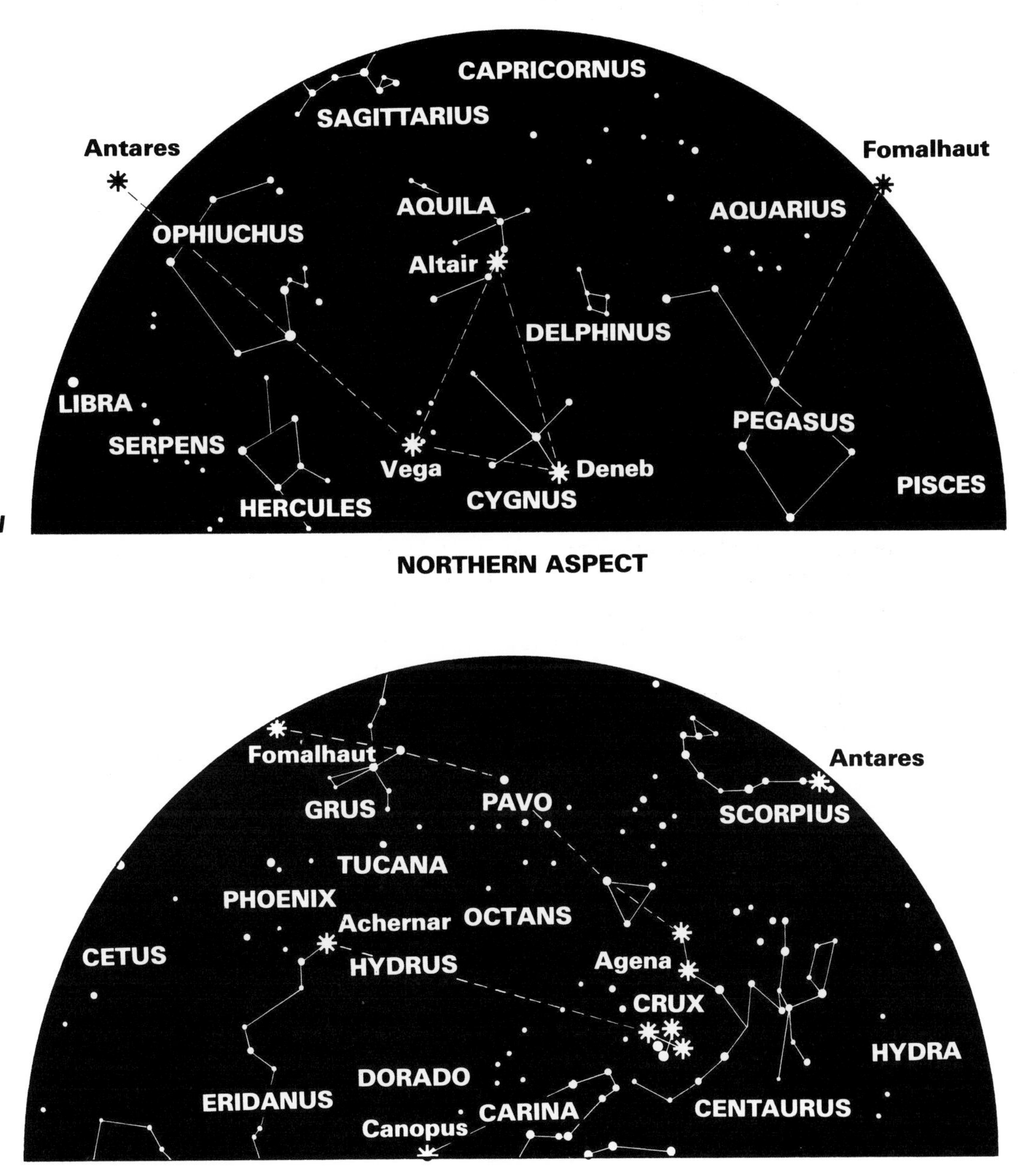
CAPRICORNUS
SAGITTARIUS
Antares
Fomalhaut
AQUILA
AQUARIUS
OPHIUCHUS
Altair
DELPHINUS
LIBRA
PEGASUS
SERPENS
Vega
Deneb
PISCES
HERCULES
CYGNUS
W
E
NORTHERN ASPECT
Fomalhaut
Antares
GRUS
PAVO
SCORPIUS
TUCANA
PHOENIX
Achernar
OCTANS
CETUS
HYDRUS
Agena
CRUX
HYDRA
DORADO
ERIDANUS
CARINA
CENTAURUS
Canopus
E
W
SOUTHERN ASPECT

28

Southern stars: the September sky

Spring is on the way, and though Scorpius is still dominant it has started to sink towards the south-west. The maps apply to:

1 June:	4 a.m.
1 July:	2 a.m.
1 August:	midnight
1 September:	10 p.m.
1 October:	8 p.m.

This is the best time of the year to see the Vega–Altair–Deneb triangle. All three are above the horizon in the north; Deneb is as high as it can ever become, though it is still rather low down. However, the richness of the Milky Way in the Cygnus region is very obvious, and the dark rifts are easy to locate. Delphinus too is showing up, and the small, compact group is quite distinctive. The 'Southern Birds' (Grus, Pavo, Phoenix, Tucana) are high, though only Grus is really striking. In the east, look out for the Square of Pegasus, which is now rising. The south-eastern area is largely occupied by Cetus. Take care not to confuse Beta Ceti with Fomalhaut, which is only too easy to do even though Fomalhaut is higher up and very much the brighter of the two. The large, faint Zodiacal constellations of Capricornus, Aquarius and Pisces are well placed, so that if you see a brilliant object in that part of the sky it can only be a planet.

STAR MAP 21

35° S 15 Sep 10 p.m.

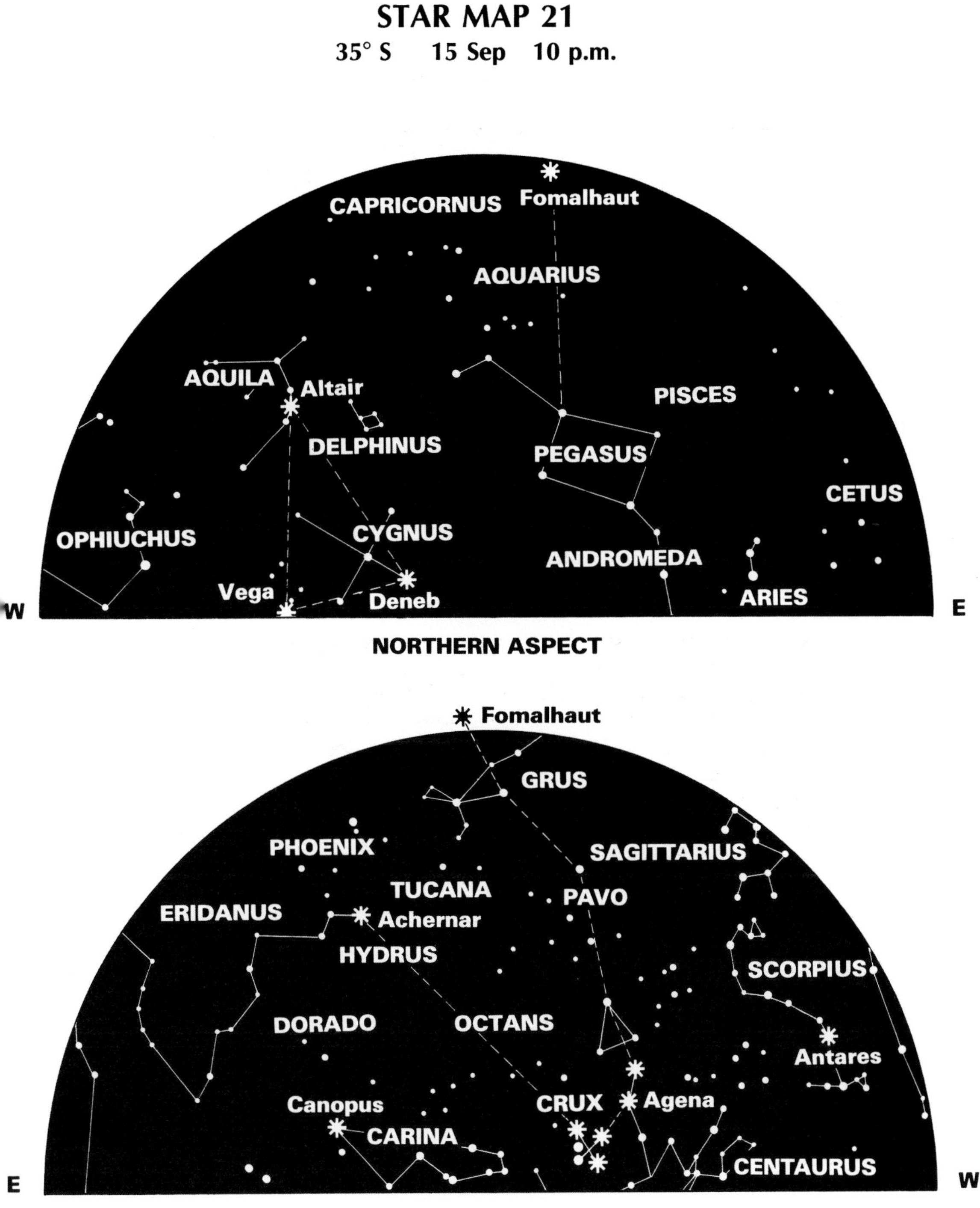

29

Southern stars: the October sky

During October evenings the night sky is probably less brilliant than at any other time of the year. These maps are for:

1 July: 4 a.m.
1 August: 2 a.m.
1 September: midnight
1 October: 10 p.m.
1 November: 8 p.m.

Scorpius is now far past its best, though it is still on view in the south-west, and Sagittarius remains fairly well placed. The most notable absentee is the Southern Cross, which is very low in the south and from some parts of South America, Australia and South Africa actually sets for a brief period. This means, of course, that Achernar is high up, while Fomalhaut is almost at the zenith. We have virtually lost the inappropriately nicknamed 'Summer Triangle', though Altair can still be seen and Deneb does not actually set until the late evening.

The main addition to the sky is Pegasus. The Square is almost due north and is at its best, so that its stars can be used as guides to Fomalhaut and Beta Ceti. The Clouds of Magellan are high, and so are the Southern Birds, with Grus particularly prominent – though all in all, I have always regarded this as the most confusing region in the entire sky

PAVO: *the Peacock*

Pavo is not the most conspicuous of the Birds, but it seems best to deal with it first because its leading star, Alpha (1.9), is bright enough to be recognizable and there is a good way to find it – if you can see the Pointers, which admittedly may be rather difficult at this time of year because they are so low in the south. Locate them, and then take a line from Alpha Centauri through Alpha Trianguli Australe, continuing it for a slightly greater distance on the far side. The first bright star to be reached will be Alpha Pavonis.

STAR MAP 22
35° S 15 Oct 10 p.m.

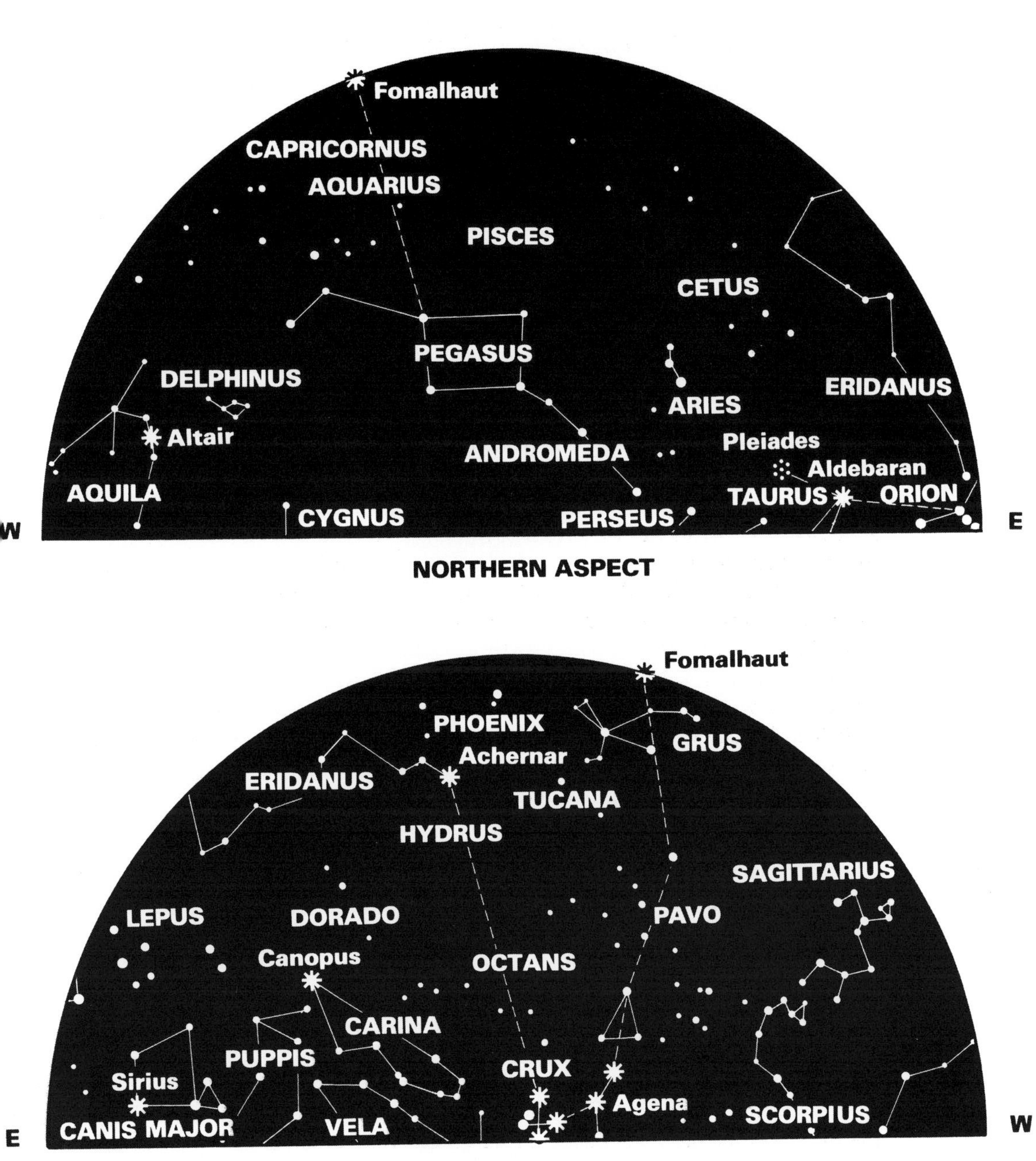
Fomalhaut
CAPRICORNUS
AQUARIUS
PISCES
CETUS
PEGASUS
DELPHINUS
ERIDANUS
ARIES
Altair
Pleiades
ANDROMEDA
Aldebaran
AQUILA
TAURUS
ORION
CYGNUS
PERSEUS
W
E
NORTHERN ASPECT
Fomalhaut
PHOENIX
GRUS
Achernar
ERIDANUS
TUCANA
HYDRUS
SAGITTARIUS
LEPUS
DORADO
PAVO
Canopus
OCTANS
CARINA
PUPPIS
CRUX
Sirius
Agena
CANIS MAJOR
VELA
SCORPIUS
E
W
SOUTHERN ASPECT

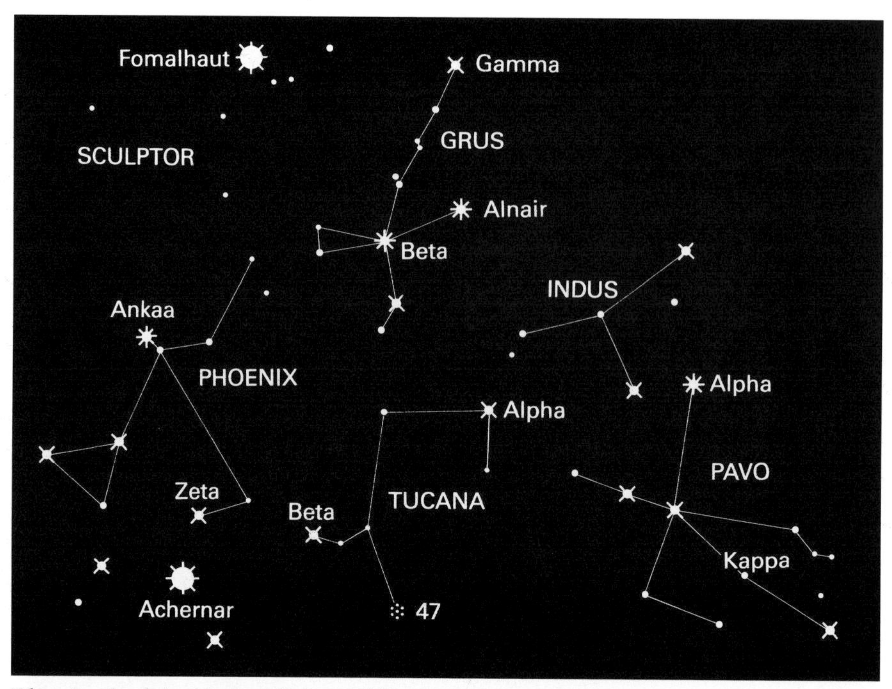

Phœnix, Sculptor, Indus, Pavo and Tucana

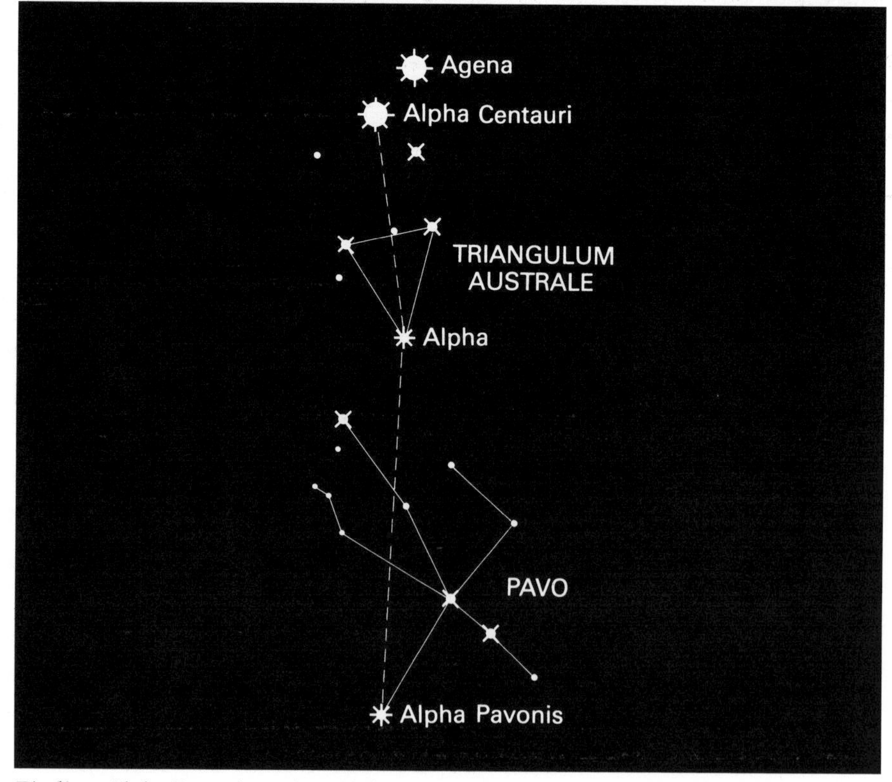

Finding Alpha Pavonis, using Alpha Centauri and Alpha Trianguli Australe

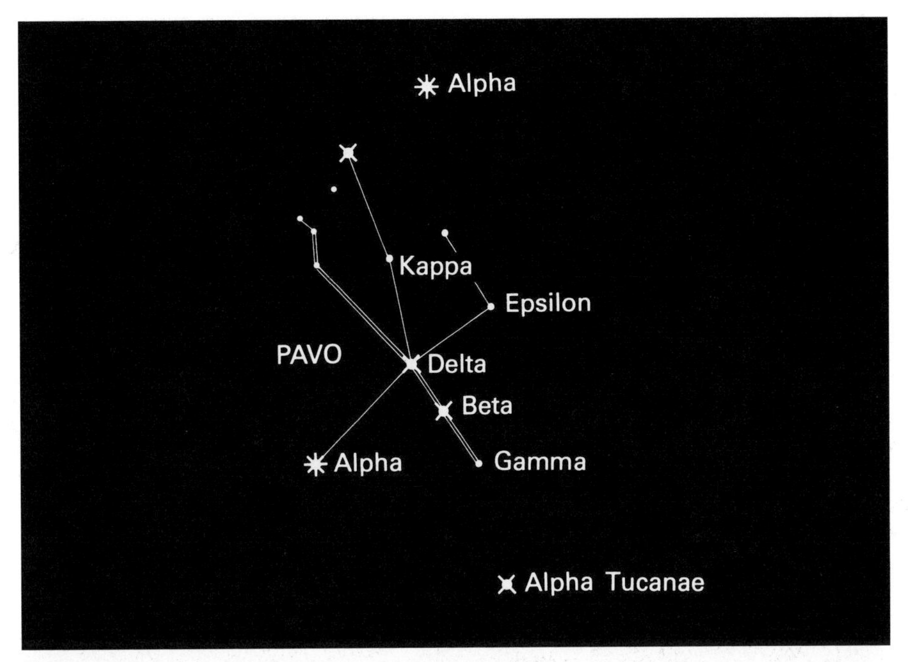

Pavo

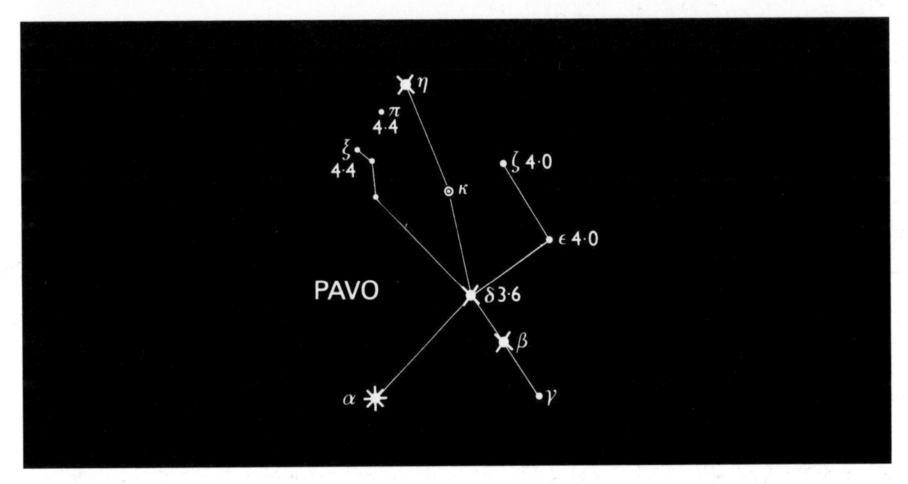

κ Pavonis

Actually, Alpha is some distance from the main part of Pavo, which is made up of a line of stars, none of which is brighter than magnitude 3.5. The most interesting naked-eye object is the short-period variable Kappa Pavonis, with a period of 9 days and a magnitude range of from 3.9 to 4.8, so that it is always visible without optical aid when the sky is clear and dark. Suitable comparison stars are Delta (3.6), Epsilon and Zeta (each 4.0), and Pi and Xi (each 4.4). Kappa Pavonis is a 'Type II' Cepheid, much less luminous than a normal Cepheid of the same period, even though its peak luminosity is much greater than that of the Sun. Its distance from us is 75 light-years.

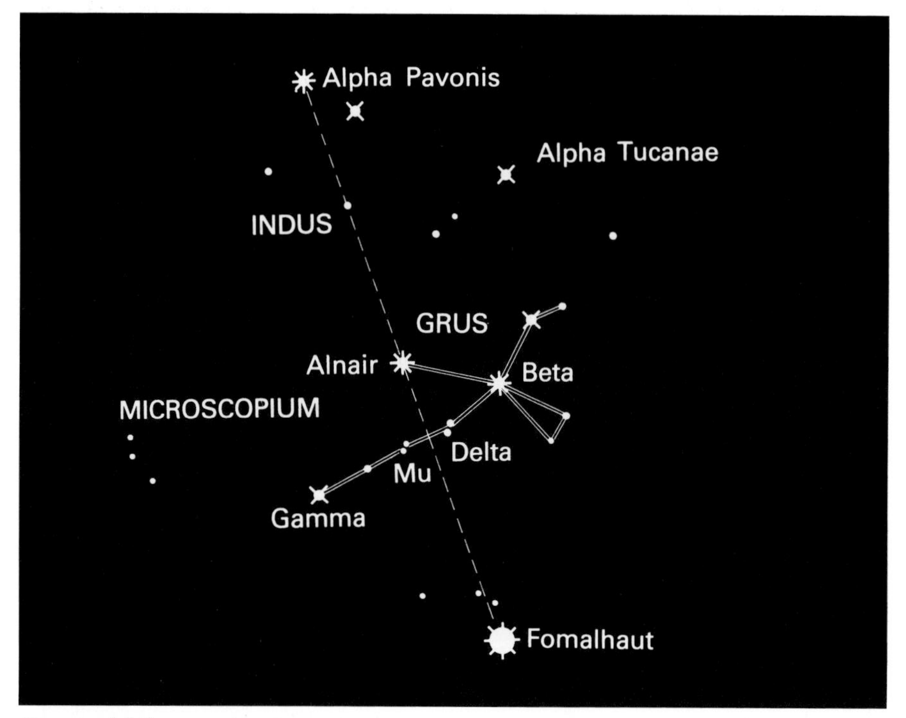

Grus and Microscopium

GRUS: *the Crane*

Grus may be said to give a vague impression of a bird in flight. It lies between Alpha Pavonis to the one side and Fomalhaut on the other, but it is in any case fairly easy to identify. Grus has two bright stars, Alpha or Alnair (1.7) and Beta (2.1). Even a casual glance will show that they are not alike: Alnair is pure white, while Beta is warm orange. Beta is, in fact, the more distant and the more powerful of the two.

The line of stars marking the main part of Grus runs from the third-magnitude Gamma through Beta to two fainter stars, Epsilon and Zeta. Between Gamma and Beta there are two pairs, Mu and Delta, which look like wide doubles but are not genuine pairs; the appearance is due merely to a line-of-sight effect.

TUCANA: *the Toucan*

Tucana lies between Achernar and Alpha Pavonis; this is probably the best way to find it. There is only one star of reasonable brightness (Alpha, 2.9), but the constellation contains part of the Small Cloud of Magellan, which extends into Hydrus. The nearest fairly prominent star to it is Beta Hydri.

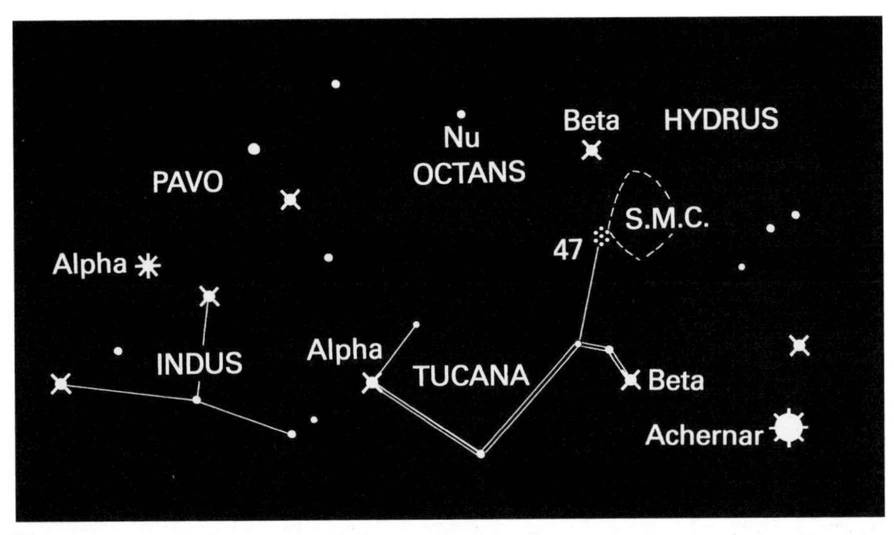

Tucana

Tucana contains two fine globular clusters. Both of them lie close to the edge of the Small Cloud, but this is sheer coincidence, because both of them belong to our Galaxy whereas the Cloud is an external system. The brighter of the two clusters, NGC 104 (C. 106), is better known as 47 Tucanae. Apart from Omega Centauri it is the brightest globular in the sky – much more prominent than M. 13 in Hercules, so that it is easily visible to the naked eye, and with a telescope it is superb. Its distance is around 16,000 light-years, about 1/12 that of the Small Cloud, and it is over 200 light-years in diameter. The other globular, NGC 362 (C. 104), is very difficult to distinguish with the naked eye, though it would be easier were it not almost in front of the Cloud. I have never been confident of seeing it without using binoculars.

Beta Tucanae is a naked-eye double, with components almost equal at magnitude 4.5. It is not easy to split the pair without optical aid, but against a dark sky it can be done.

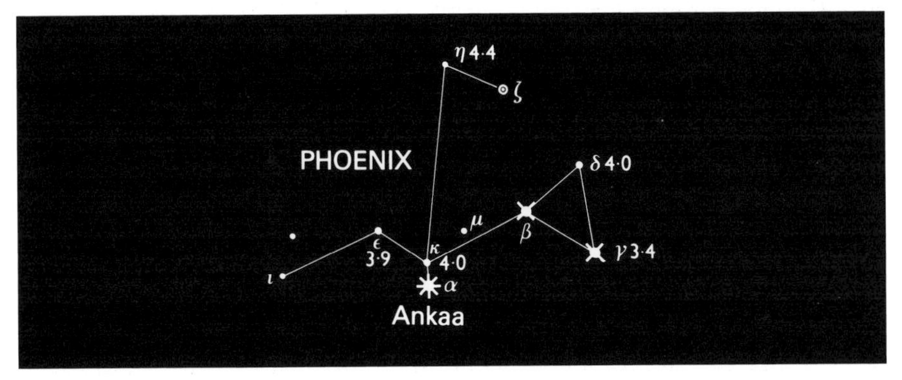

ζ Phœnicis

PHOENIX: *the Phoenix*

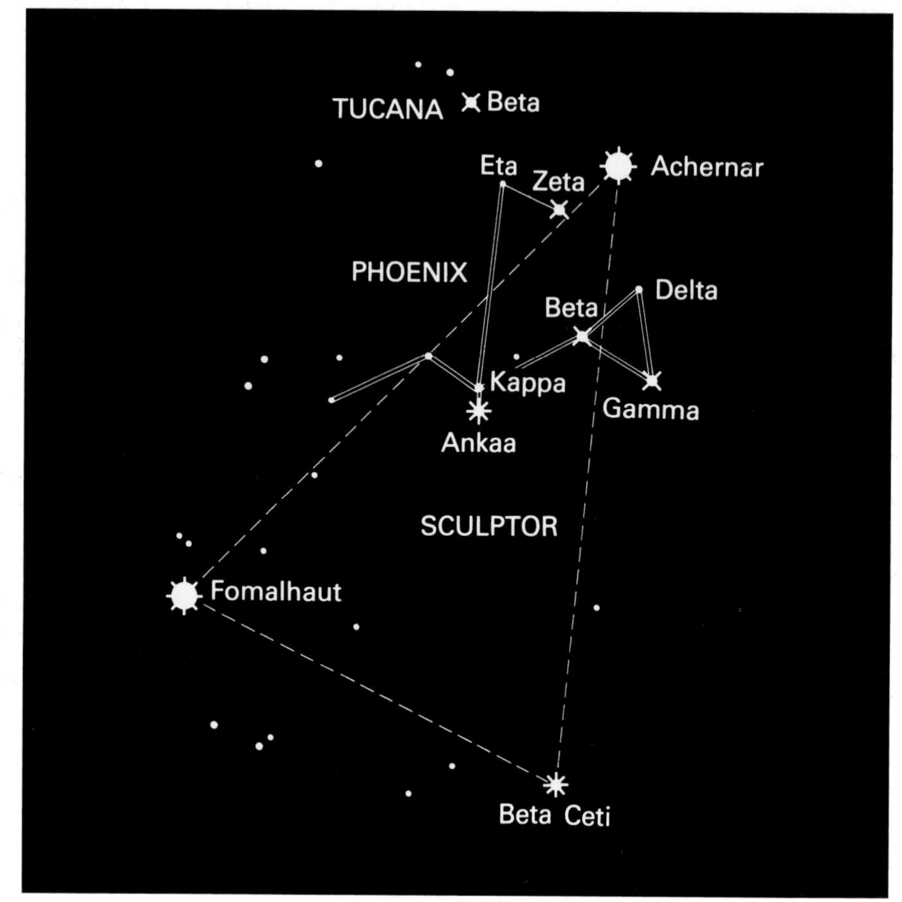

Finding Phœnix, using Beta Ceti, Fomalhaut and Achernar

Phoenix is named after the mythological bird which periodically burnt itself to ashes and was promptly reborn. It lies inside the triangle formed by Achernar, Fomalhaut and Beta Ceti, and adjoins the Crane and the Toucan; its brightest star, Alpha or Ankaa (2.4), lies between Achernar and Fomalhaut, though somewhat away from the direct line joining them. The only interesting object is Zeta, an Algol-type eclipsing binary with a period of 1.7 days and a magnitude range of between 3.6 and 4.1. Comparison stars are Gamma (3.4), Epsilon (3.9), Kappa (4.0, near Ankaa) and Eta (4.4). Zeta Phoenicis is not hard to locate, because it lies near Achernar, and under good conditions it never drops below naked-eye visibility.

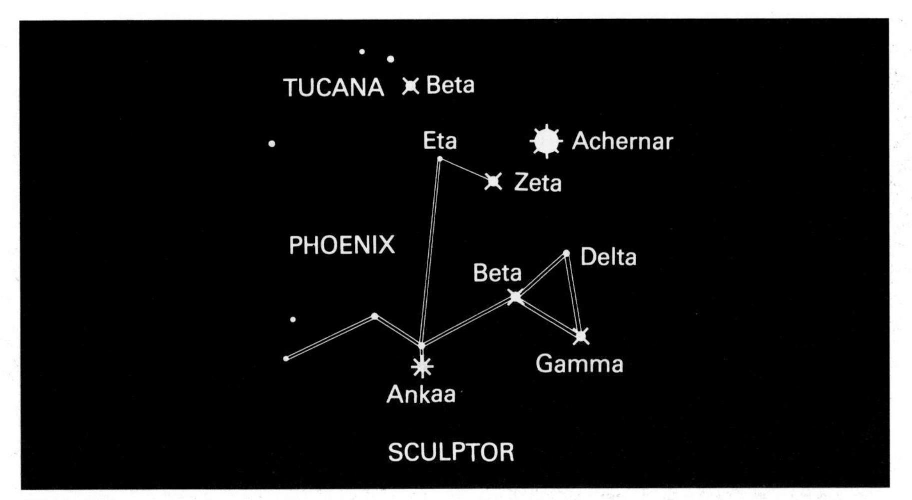

Phœnix

INDUS: *the Indian*

A dull constellation, with no notable features and only one star, Alpha (3.1), of reasonable brightness. It lies between Alpha Pavonis and Alnair.

MICROSCOPIUM: *the Microscope*

Microscopium has no star as bright as magnitude 4.5, and is entirely uninteresting. It adjoins Indus and Grus.

30

Southern stars: the November sky

With the approach of summer Orion is making its entry in the east. It is still low down, but of course it gains altitude quickly and is well displayed in the early hours of the morning. These charts are valid for:

1 August: 4 a.m.
1 September: 2 a.m.
1 October: midnight
1 November: 10 p.m.
1 December: 8 p.m.

By now Scorpius has almost vanished in the evening twilight, and the Southern Cross is too low to be properly seen; Achernar lies near the zenith, with Fomalhaut also very high. Canopus is high enough to be prominent, and Sirius is coming into view. The Square of Pegasus is in the north; extending eastward from it you will be able to make out the line of stars marking Andromeda, and this is a good time to look for the Great Spiral, M. 31, though from southern countries it is difficult to see with the naked eye and from the southern part of South America and New Zealand it will not be visible at all. The whole of Eridanus is displayed, and so is Cetus, so that if Mira is near maximum it can be found without any difficulty. This is the worst time of the year for seeing the Milky Way, which lies almost along the horizon. However, the Clouds of Magellan are excellently placed, together with those confusing Southern Birds.

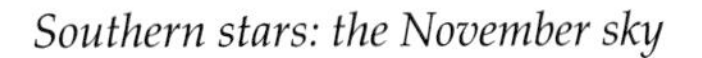

STAR MAP 23

35° S 15 Nov 10 p.m.

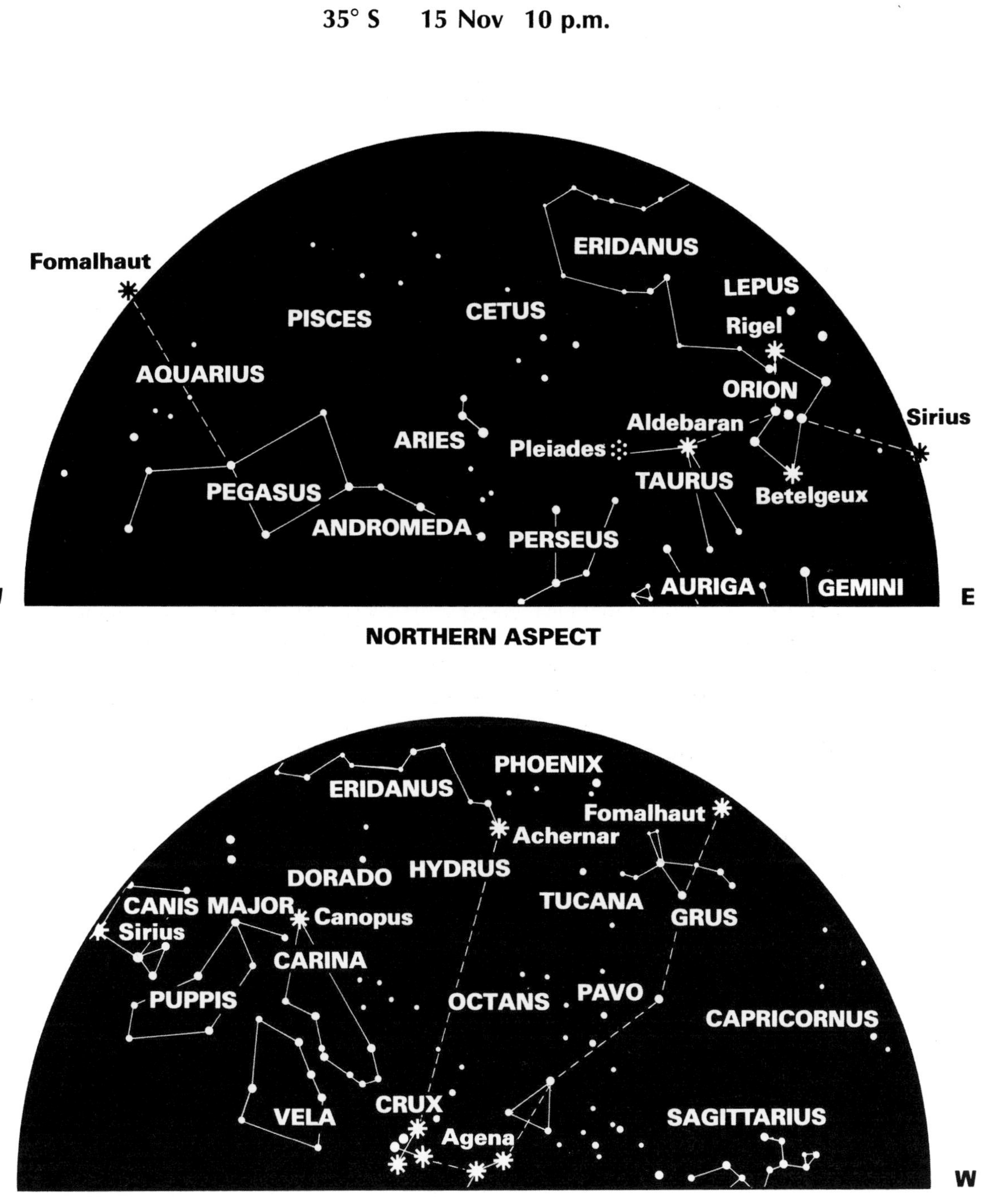

31

Southern stars: the December sky

Finally, in our seasonal review of the sky, we arrive at December, and the summer stars are back. The charts are for:

1 September: 4 a.m.
1 October: 2 a.m.
1 November: midnight
1 December: 10 p.m.
1 January: 8 p.m.

Orion is high enough to be useful once more as a guide, and it shows the way to the whole of its retinue apart from Procyon, which has barely risen, and the Twins, which are still below the horizon. Sirius is glorious; so, higher still, is Canopus. Achernar is almost overhead, Fomalhaut to the south-west of the zenith, and the Clouds of Magellan are at their best, though the Southern Cross is still very low and will not be properly seen until the early hours of the morning. Pegasus is setting, but Andromeda is still skirting the northern horizon, and you will be able to make out Aries and Triangulum. This is also a good time to look for Algol in Perseus, and despite its low altitude the minima can be followed without difficulty.

STAR MAP 24

35° S 15 Dec 10 p.m.

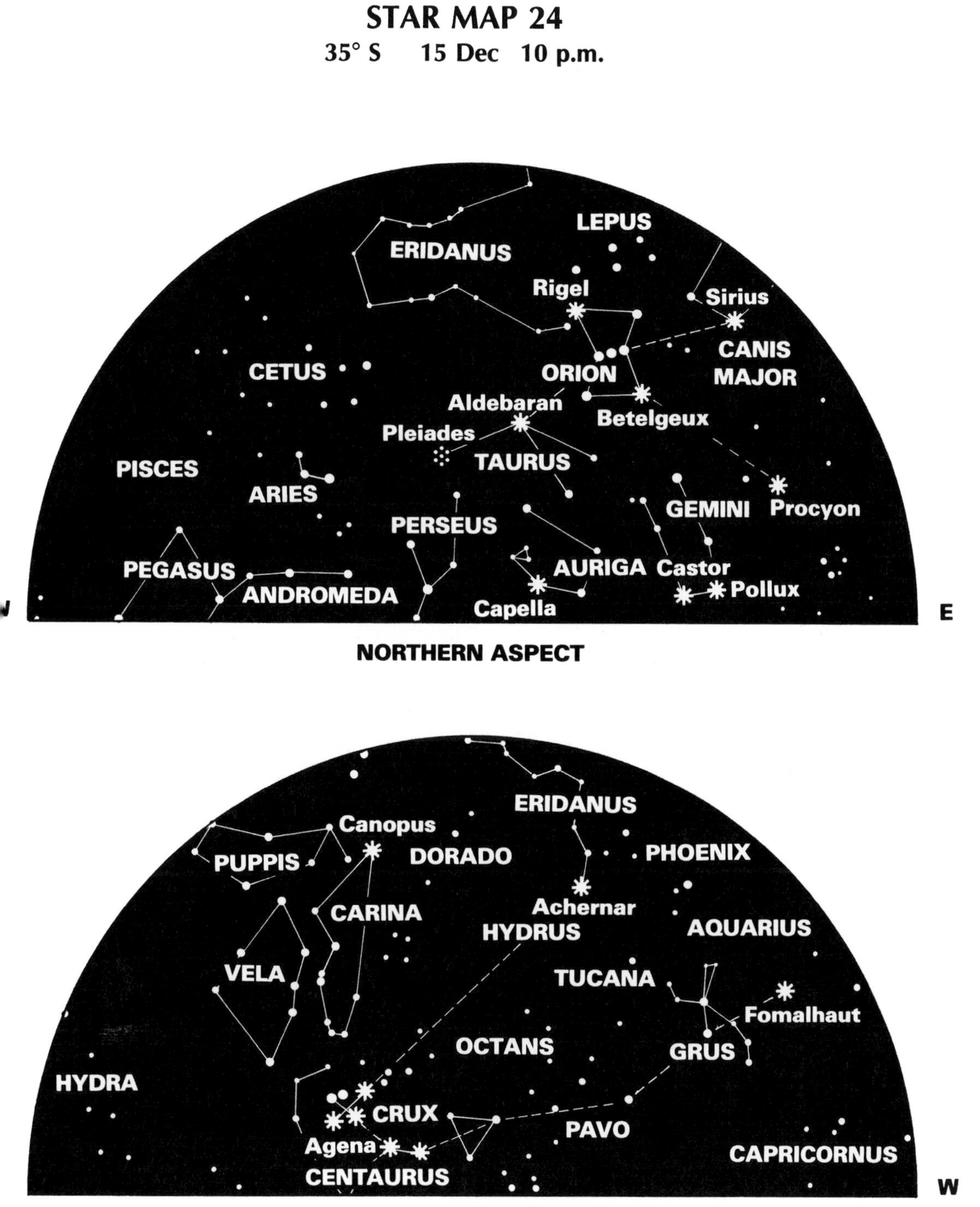

32

The equatorial sky

Some years ago I paid my first visit to Manila, in the Philippine Islands. The latitude there is less than 15 degrees north, and I found the view of the night sky unfamiliar; both Ursa Major and the Southern Cross were quite high above the horizon.

From the equator, the two celestial poles lie on opposite horizons, and the celestial equator passes directly overhead. No star is circumpolar (not even Polaris, because it is almost a degree away from the actual pole and so describes a very small daily circle; if you observe from latitude 0°, Polaris will in theory bob up and down over the skyline, though things are admittedly complicated by the effects of refraction). On the other hand, the entire heavens are available at some time or another. And of course there are no 'seasons'.

The regions between latitudes 15 degrees north and south include parts of South America, parts of Africa, the East Indies, the Philippines and the extreme northern tip of Australia. Singapore is a mere 1° 20′ north. Very few books cater for people who live in these parts of the world, so it may be useful to give some maps here. Obviously, allowance must be made for your latitude; if you are more than 1 degree south you will never see Polaris, and if you are more than 1 degree north you will never see Sigma Octantis (though this would be difficult in any case, because it is so faint that any horizon glow or haze will hide it). But these charts do apply to those observers who are fairly close to the equator.

THE JANUARY SKY

The charts for 10 p.m. in January also apply to around midnight in December, 2 a.m. in November, 8 p.m. in February and 6 p.m. in March. The first thing to note is that Orion is overhead, so that it has to be split between the two hemispherical maps; Betelgeux is in the northern aspect and Rigel in the southern. Capella is high in the north, with Procyon and the Twins; Leo is rising in the north-east, Cassiopeia very low in the north-west. A few stars of Ursa Major can be seen, but the Plough (or Dipper) itself is mainly below the horizon.

Sirius and Canopus are dominant in the south, and the richness of the Milky Way is very evident. Both the south-east (occupied largely by Hydra) and the south-west (Cetus and Eridanus) are rather barren. Achernar, the only bright star in Eridanus, is well above the horizon, which means of course that the Southern Cross is absent.

STAR MAP 25
EQUATOR 15 Jan 10 p.m.

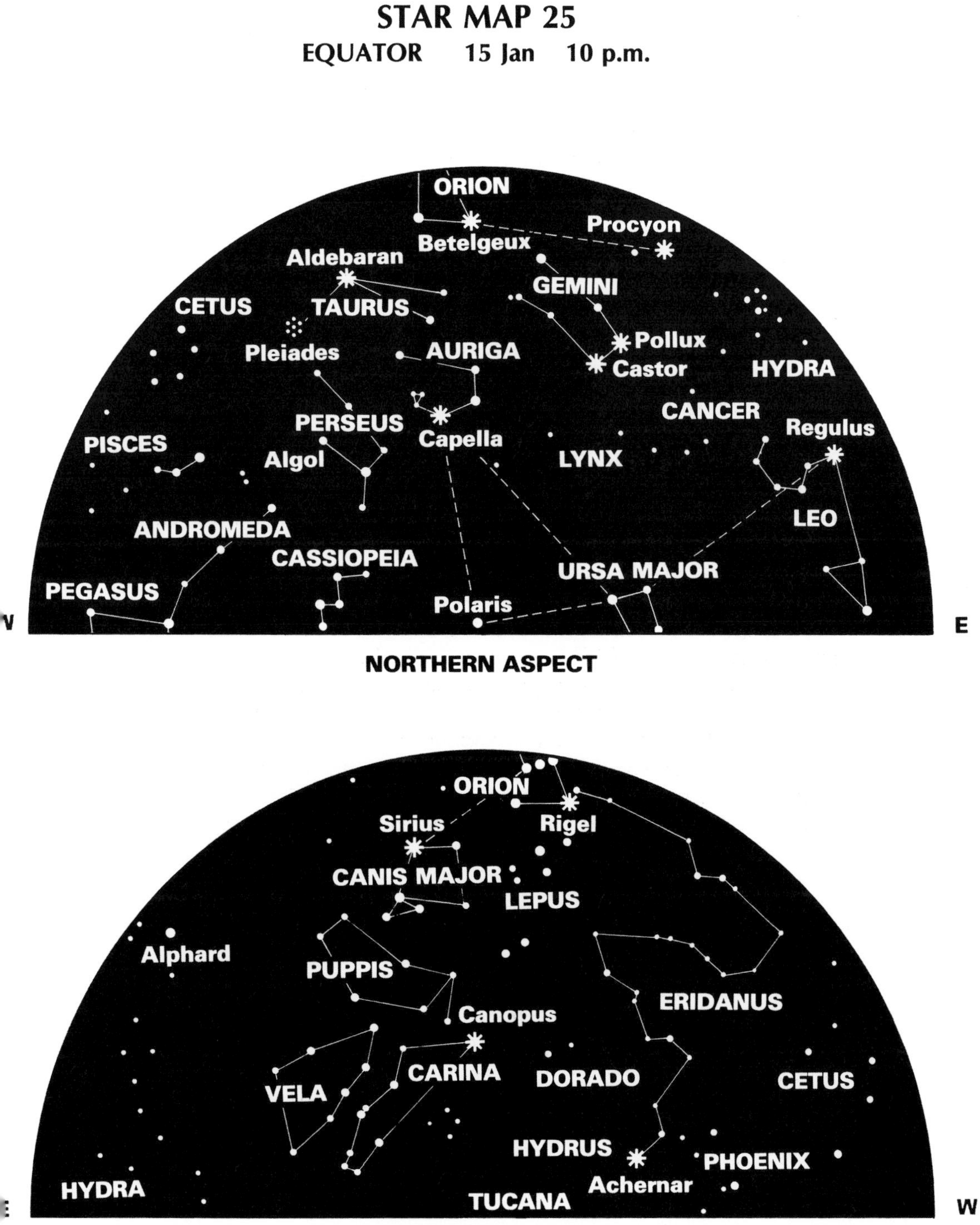

THE APRIL SKY

The charts for 10 p.m. in April (4 a.m. in January, 2 a.m. in February, 8 p.m. in May and 6 p.m. in June) are the first to show Ursa Major and the Southern Cross at the same time. Moreover, they are at roughly equal heights above the horizon, so that they are practically opposite each other. Orion has almost disappeared.

There are not too many brilliant stars in the northern map, though Regulus in Leo and Arcturus in Boötes are high up; Procyon and the Twins are descending in the west, while the barren Hercules region is rising in the east. Now that Ursa Major is visible, it is a good time to look for Polaris. If you are even marginally north of the Earth's equator you should be able to glimpse it under really good conditions, but if you are at latitude 0° I doubt if you will have any success, and of course from 1° south it never rises at all.

Antares is rising in the south-east, while Sirius has almost disappeared in the south-west. Of course the whole of the southern aspect is dominated by Centaurus and Crux, even though we have lost Canopus. Also, note the quadrilateral of Corvus very high in the south. Though its stars are not really bright, the shape of the constellation makes it very easy to identify. The only conspicuous star in the south-west area is Alphard in Hydra.

THE JULY SKY

These charts are for 10 p.m. in July, which means 4 a.m. in April, 2 a.m. in May, midnight in June, 8 p.m. in August and 6 p.m. in September. Almost all Ursa Major has set (you may still be able to find Alkaid) and so has the Southern Cross; Alpha and Beta Centaur do remain above the horizon, but they are very low slightly west of south. In the northern aspect the Vega–Deneb–Altair triangle is prominent, with Arcturus north-west of the zenith; in the southern aspect the Scorpion is dominant, with Fomalhaut low in the east and Spica low in the west. Now it is the overhead region which is barren, because it is occupied by the Hercules–Ophiuchus group of large, dim constellations. The Milky Way is magnificent, because two of the richest regions (Cygnus in the north, Scorpius–Sagittarius in the south) are both high up.

THE OCTOBER SKY

Our last pair of maps applies for 10 p.m. in October, and for 4 a.m. in July, 2 a.m. in August, midnight in September, 8 p.m. in November and 6 p.m. in December. In the northern chart Pegasus is at its best, and is high up, while the Vega–Deneb–Altair triangle is descending in the west; Vega has become very low. Cassiopeia is well above the northern horizon, and this is also the

STAR MAP 26

EQUATOR 15 Apr 10 p.m.

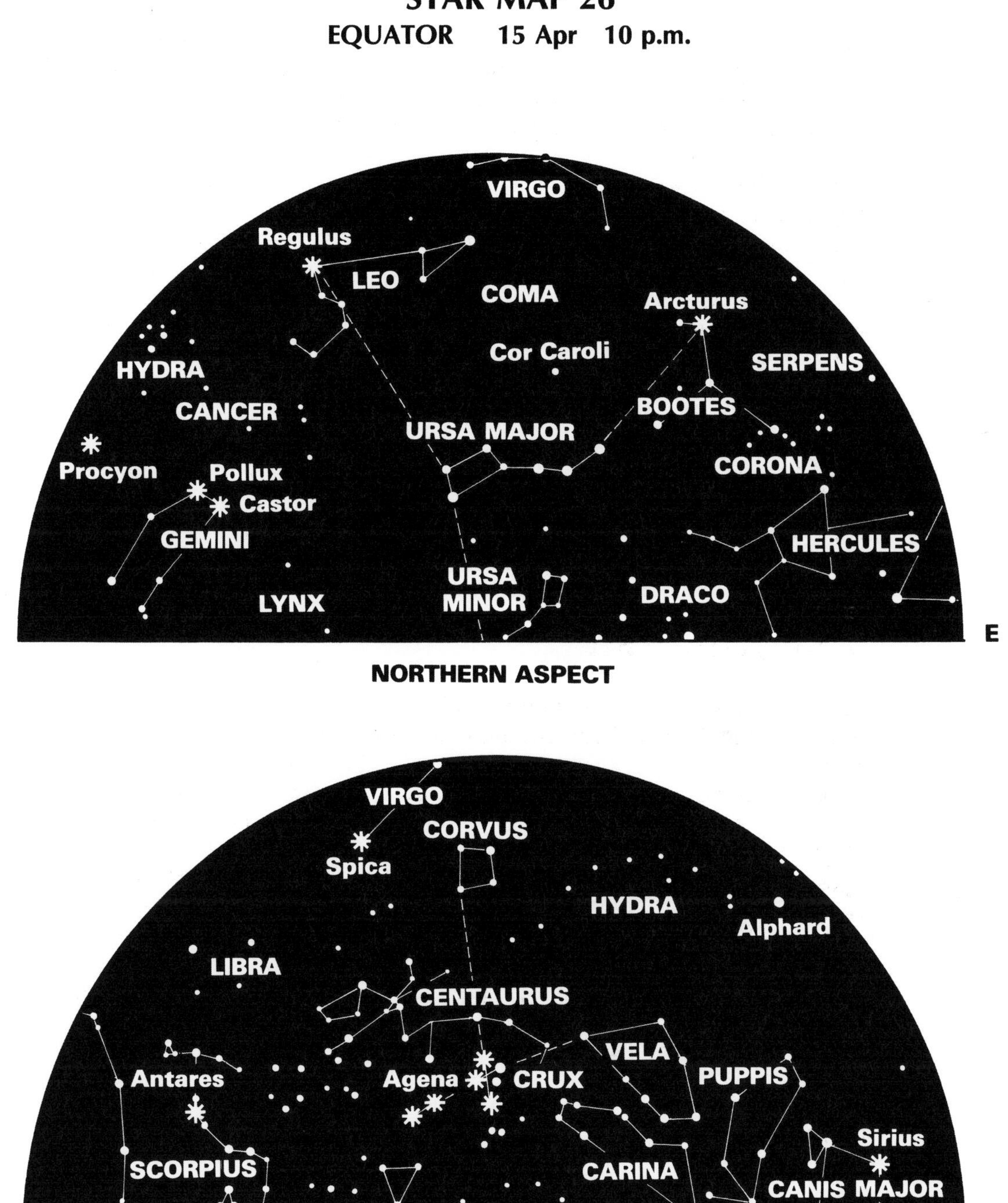

STAR MAP 27

EQUATOR 15 Jul 10 p.m.

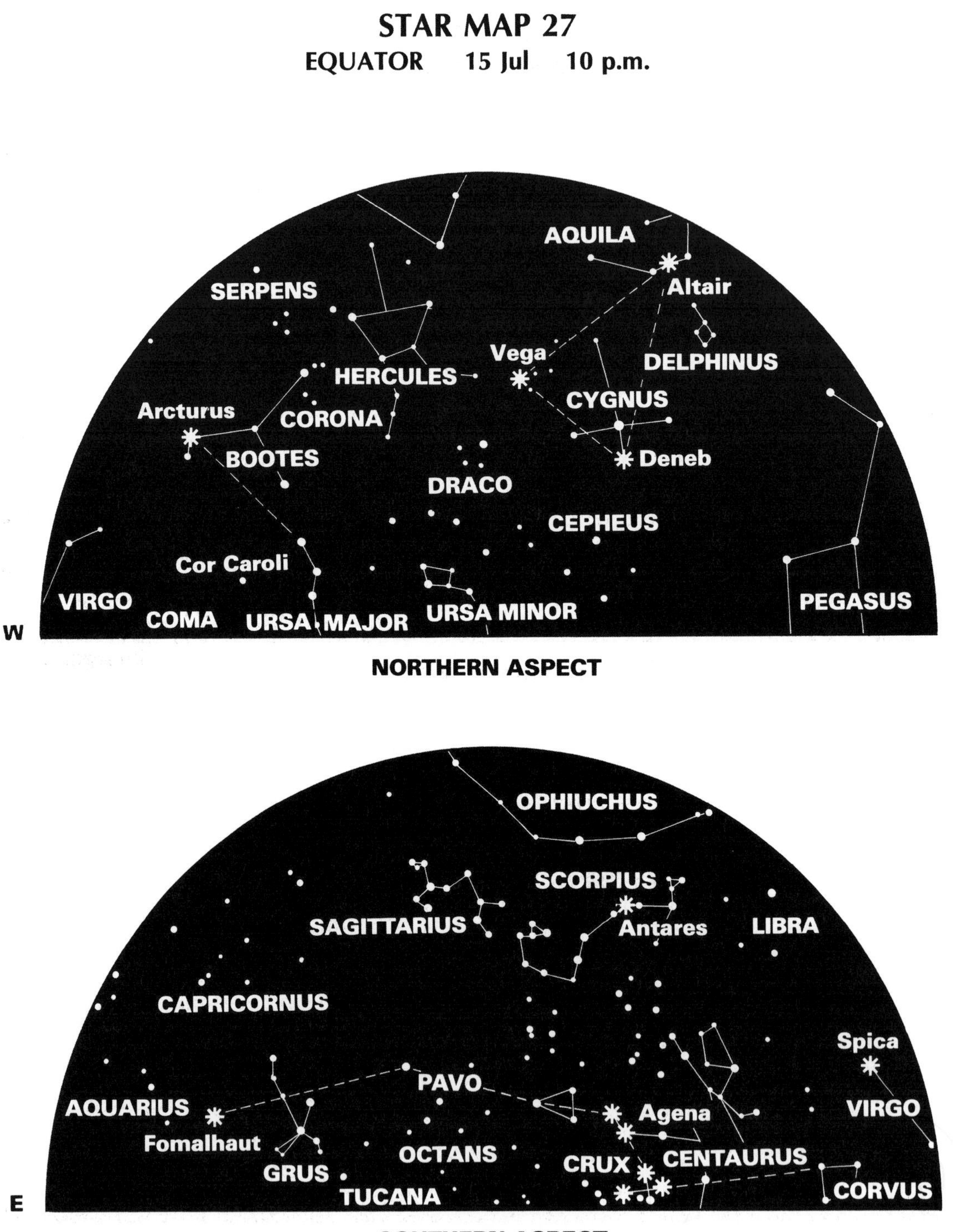

STAR MAP 28

EQUATOR 15 Oct 10 p.m.

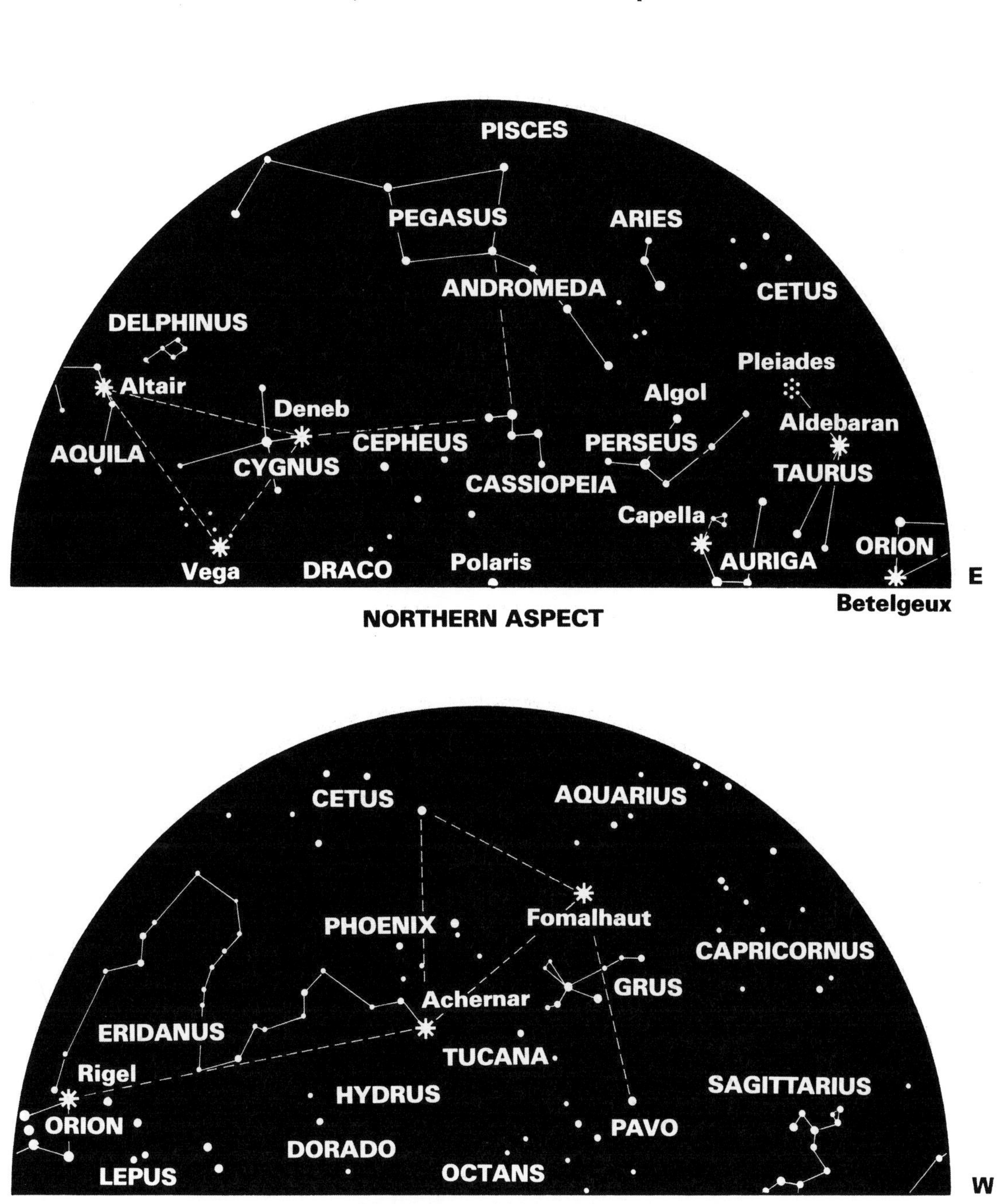

best time to find Cepheus, with its two famous variable stars Delta and Mu. Capella is rising, and you may be able to see part of Orion; Betelgeux grazes the horizon, but in the southern map, to the east, Rigel has risen. Achernar too is back, but we have lost Scorpius, and Sagittarius will soon follow it below the horizon. Fomalhaut is not far from the zenith, and below it you will easily find the distinctive shape of Grus, after which you should be able to locate the other Southern Birds. Again the overhead region is barren, because it is filled by the dim groups of Aquarius, Pisces and Cetus.

This brings us to the end of our brief survey of the stellar sky. I hope that it will help you to find your way around, wherever you may live.

33

Our star: the Sun

We owe everything to the Sun. Without it we could not exist, and the Earth itself would never have come into being. I have already said something about the way in which the Sun shines, and how it evolves, so let me now turn to what the naked-eye observer can see.

The answer is – not much! Occasionally sunspots large enough to be seen with the naked eye come into view, but I do not advise looking for them, or even looking at them when you know that they are present. The Sun is extremely dangerous. I have been accused of over-stressing the risks, but I am quite unrepentant, because I was 'warned off' when I was very young; I met an old amateur astronomer who was blind in one eye. He had been so ever since the age of nine, when he was looking at the Sun through a telescope equipped with a dark filter over the eyepiece, and the filter cracked so suddenly that he was unable to remove his eye from the danger-zone before the sight in it had been permanently destroyed.

Therefore, there is only one rule for looking at the Sun through any telescope or binoculars, even when a dark filter is fitted: *Don't.*

When low down in the sky and shining through a layer of haze, the Sun looks innocent and harmless enough – but it isn't. Even then I do not recommend staring at it to see whether there are any sunspots around. It is true that when the Sun is very low, there is no real harm in looking at it with the naked eye if you use a dark glass, but the glass must be really dark. You will probably see that the Sun looks flattened. This is due to the Earth's atmosphere, which bends or refracts the sunlight. The closer the object is to the horizon, the greater the amount of refraction; the Sun's lower limb is more strongly refracted than the upper. Refraction also tends to lift up anything which is near the horizon, and there are a few cases on record when the Sun and the full moon have been seen simultaneously at opposite horizons, though one or the other must theoretically be out of view.

Occasionally the last segment of the Sun's disk flashes brilliant green before it disappears. This Green Flash is said to be very striking, and it has been photographed often. It too is due to effects of the Earth's atmosphere. I have seen it only once, and then not well. The best hope is to look for it against a sea horizon – but, I repeat, never use any telescope or even binoculars.

Now let us turn to eclipses of the Sun.

The theory is straightforward enough. As the Moon revolves round the Earth while the Earth revolves round the Sun, there must be times when the

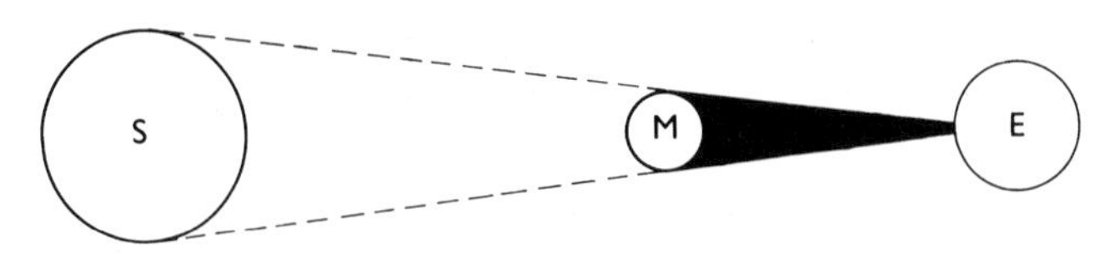

Theory of solar eclipse. S – Sun; M – Moon; E – Earth. The diagram is not to scale.

three bodies are exactly lined up. By sheer coincidence the Sun and the Moon appear to be virtually the same size in the sky, with diameters of about half a degree each, so that the Moon is just able to cover the Sun.

The diagram shows what happens. The cone of shadow cast by the Moon is only just long enough to reach the Earth's surface. When the eclipse is total, so that the bright solar surface is completely hidden, the zone of totality can never be more than 170 miles wide. Also, no total eclipse can last for as long as eight minutes, and most are much shorter, so that you have to be in exactly the right place at exactly the right time. From England, the last total eclipse took place on 11 August 1999, when the zone of totality crossed Cornwall. The next English total solar eclipse will not be until 2090 – again in the West Country.

On average there are two or three solar eclipses each year, but by no means all of them are total, and the eclipse will be seen as 'partial' to either side of the belt of totality; moreover, many eclipses are not total from anywhere on the Earth. There is also another complication. The Moon's distance from the Earth is not constant, and when it is at its furthest from us it appears slightly smaller than the Sun, so that when the lining-up is perfect a ring of sunlight is left showing round the dark disk of the Moon. This is called an annular eclipse (from the Latin *annulus*, a ring).

Solar eclipses between 2000 and 2005 are as follows:

2000	5 February	Partial (56%)	Antarctic
2000	31 July	Partial (60%)	Arctic
2000	25 December	Partial (62%)	Arctic
2001	21 June	Total (4m 56s)	Atlantic, South Africa
2001	14 December	Annular (3m 54s)	Central America, Pacific
2002	10 June	Annular (1m 13s)	Pacific
2002	4 December	Total (2m 4s)	S. Africa, Indian Ocean, Australia
2003	31 May	Annular (3m 36s)	Iceland, N. Scotland
2003	23 November	Total (1m 57s)	Antarctic
2004	19 April	Partial (74%)	Antarctic
2004	14 October	Partial (93%)	Arctic
2005	8 April	Total (0m 42s)	Pacific, America
2005	3 October	Annular (7m 9s)	Spain, Africa, Indian Ocean

It is obvious that a solar eclipse can happen only at new moon, but less obvious why eclipses do not happen every month. The reason is that the Moon's orbit is tilted to ours at an angle of just over five degrees, so that at most new moons there is no lining-up and therefore no eclipse; the Moon passes unseen either above or below the Sun in the sky.

A total eclipse is magnificent because as soon as the last segment of the bright disk is hidden, the dark mass of the Moon is seen to be surrounded by the pearly corona. The corona has nothing to do with the Moon; it is the Sun's outer atmosphere, and is made up of very thin gas, mainly hydrogen. Because it is relatively faint, it is quite invisible with the naked eye except during totality; neither will any ordinary telescope show it. The corona stretches outward from the Sun in all directions, but it has no sharp boundary; it simply thins out until it becomes undetectable.

Below the corona comes the chromosphere or 'colour-sphere', marking the lower atmosphere of the Sun. It is here that we find the prominences, which look red but are certainly not flames. They are made up of incandescent hydrogen, and are truly vast: the length of an average prominence is around 120,000 miles.

Modern instruments based upon the principle of the spectroscope make it possible to study the prominences at any time, without waiting for an eclipse, but it is only during totality that they can be seen with the naked eye. Sometimes they are very striking, but unfortunately it is essential for the whole of the Sun's disk to be covered up by the Moon in order to see them; 99 per cent will not do. Only when the last sliver of the bright surface has been hidden do the chromosphere, the prominences and the corona leap into view. Just as the eclipse begins there is the wonderful 'diamond ring' effect, as the remaining fraction of the Sun shines out from behind the advancing Moon. It lasts for only a second or two, but it is one of the most beautiful sights in all nature; and it is seen briefly again as totality ends. When totality is just starting it is also possible to see what are called 'Baily's Beads', caused by sunlight shining through valleys at the Moon's edge.

Quite apart from the glory of a total eclipse, some useful observations can be made. It is worth keeping track of the changes in wind velocity, barometric pressure and temperature; the drop is very noticeable. I well remember that in Java, in 1983, the pre-eclipse temperature was furnace-like at well over 100 degrees Fahrenheit, and the coolness during the eclipse period came as a welcome relief! Then there are the curious wavy shadow-bands which may be seen crossing the landscape just before and just after totality; they are not always visible, but they are worth photographing.

At mid-totality the brightest stars and planets are clearly visible. At the Caribbean eclipse of 1998 I could see Venus for at least twenty minutes before totality and nearly half an hour afterwards, and first-magnitude stars shone out. No two eclipses are alike, however; in 1968, when totality crossed the Yurgamysh region of Siberia, the sky was so bright that I failed to see any stars at all.

Various somewhat 'off-beat' investigations have been made from time to time. At one eclipse in 1878, a serious search was made for a planet which was thought to move round the Sun at a distance less than that of Mercury. It had even been given a name, Vulcan, and the great French astronomer Le Verrier (who had died in the previous year) was quite convinced of its existence. Two Americans, Swift and Watson, sketched the stars which showed up round the eclipsed Sun, and claimed that they had found not one new planet, but several. Unfortunately it is now certain that they had recorded nothing more significant than ordinary stars, and Vulcan does not exist.

On the other hand, comets are always apt to take us by surprise. During the 1882 eclipse a comet was photographed during totality, when it was very near the Sun; it had never been seen before and was never seen again, so that this is our only record of it. Another eclipse comet was seen in 1948, though it was subsequently tracked. The chances of finding an unexpected comet during totality are slight, but not nil. I have been lucky enough to see eight totalities, and each time I have searched for comets, but with no success.

Glorious photographs of the corona are taken at every total eclipse when clouds do not interfere, and an ordinary camera, set on a firm tripod, is quite adequate. But again let me stress the risks. An SLR (single-lens reflex) camera is a trap, because looking at the Sun in order to focus is dangerous – it is wiser to line up by using the camera's shadow on a screen. There are various materials which can be fitted across the lens to make direct sighting safe, but it is essential to take the greatest care. During actual totality it is quite safe to look directly at the Sun, because the light-level is no more than that of the full moon, but be sure to take your eye away from the camera lens well before totality is due to end. Leave nothing to chance. Of course, splendid drawings can be made by anyone with artistic ability (something which I completely lack).

Eclipse stories are legion. I will pass quickly over the experience of one observer who travelled five thousand miles to see an eclipse lasting for five minutes, and took forty pictures of the corona under ideal conditions – but with the lens-hood of his camera in position the whole time ... And on many occasions isolated clouds have covered the Sun for the vital minutes. To see a total eclipse, you need luck as well as foresight.

Partial or even annular eclipses are tame by comparison, and the naked-eye observer can do no more than note the increasing 'bite' out of the Sun as the Moon advances. But totality is breathtaking, and if you have any opportunity of seeing a total eclipse I strongly recommend you to take it.

34

The craters of the Moon

Just as the Sun dominates the daytime sky, so the Moon is pre-eminent at night. When the Moon is full it is brilliant, but looking directly at it, either with the naked eye or a telescope, is safe enough, because the amount of heat which we receive from it is negligible. You may dazzle yourself, but you cannot damage your eyesight.

The Moon's diameter is only 2160 miles; its distance from us ranges between 221,000 miles and 253,000 miles. Oddly enough it is not very reflective, and sends back only about 7 per cent of the sunlight which it receives, so that the lunar rocks are decidedly dark. During the crescent stage, the 'night' side can usually be seen clearly, producing the effect often called 'the Old Moon in the New Moon's arms'. This is due to light reflected on to the Moon from the Earth.

If you check the position of the Moon against the stars from night to night, you will see that it is moving eastward. It covers about 13 degrees in 24 hours, so that in one hour it shifts by about half a degree – the same distance as its own apparent diameter. Obviously, then, it will rise later every night. The 'retardation', or delay in moonrise from one night to the next, is generally more than three-quarters of an hour, but the situation in September is rather different, because during autumn in the northern hemisphere the Moon's path in the sky makes a comparatively shallow angle with the horizon. This is the time of Harvest Moon.

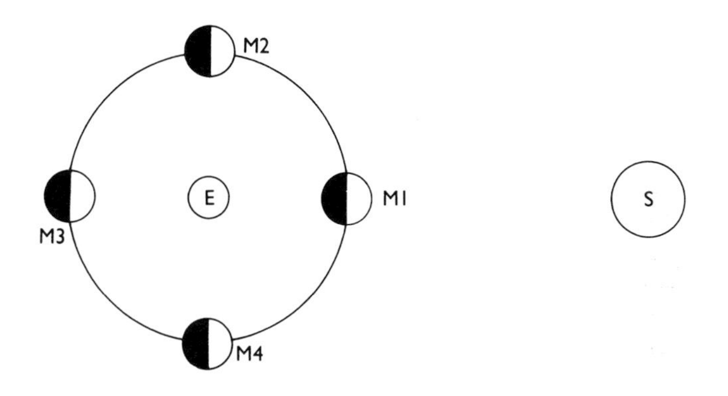

Phases of the Moon. S – Sun; E – Earth; M1 to M4 – the Moon in four different positions in its orbit. Not to scale.

Look at the diagram below, which shows the angle of the ecliptic in northern spring and autumn. (The Moon's path does not lie exactly along the ecliptic, but it is not very different.) In spring, the Moon moves from A to B in 24 hours, so that it must rise much later by the time it has reached B. In autumn, the Moon moves the same distance (C to D) in 24 hours, but the delay in the time of rising may be reduced to little more than a quarter of an hour. The September full moon will shine brightly in the late evening sky for several nights running, though it is not true to say, as many books do, that the Moon rises at almost the same time for a week or so. The following full moon is called Hunter's Moon.

It has also been said that the Harvest Moon looks particularly big, and people have described it as being 'the size of a dinner-plate'. Actually, Harvest Moon is no larger than any other full moon, and in any case the Moon is a surprisingly small object in the sky. A disk one-fifth of an inch in diameter held at arm's length will cover it. If you doubt me, cut out a disk of that size, and try the experiment for yourself.

Moreover, to describe the Moon as being the size of a dinner-plate is about as useful as saying that it looks as big as a piece of wood. To measure the apparent diameter of a celestial object in feet or inches is impossible; you have to use angular measure. Artists who paint moonlit scenes nearly always exaggerate the size of the Moon, though I suppose that this is permissible on grounds of artistic licence.

Another popular misconception is that the full moon looks largest when it is low down, and shrinks as it rises above the horizon. This celebrated 'Moon Illusion' has been known for two thousand years at least, and has led to endless arguments.

Nobody doubts the illusion; to the casual observer, a low-down full moon seems enormous. But measure it, using your one-fifth inch disk, and you will find that it *is* an illusion and nothing more. When low, the full moon is no larger than it is when riding high in the sky. The obvious explanation is that the observer is unconsciously comparing the Moon with objects along the

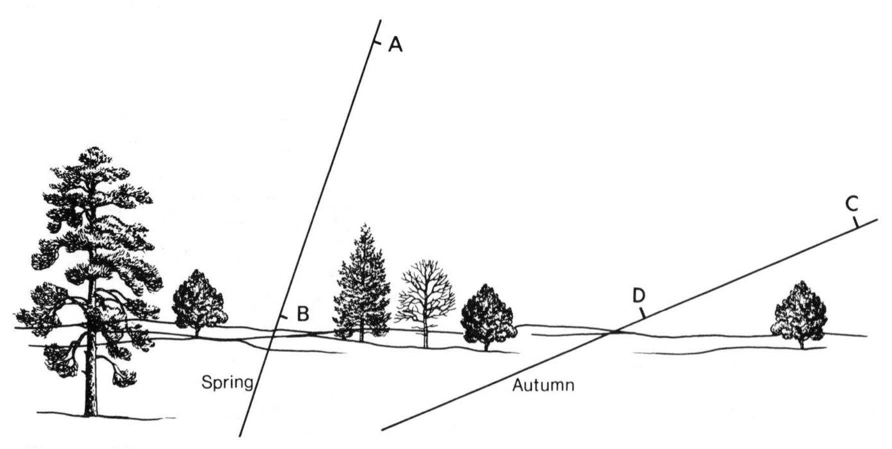

Harvest Moon.

horizon such as trees and houses, but it seems that the explanation is less simple than this. The illusion is still marked when the Moon is rising over a sea horizon, where there are no adjacent objects to act as comparisons.

It has also been suggested that the illusion is due to the mechanism of the human eye, so that an object appears larger when viewed horizontally. Lying on one's back and observing from a supine position shows that this idea will not do either, and nowadays most people have come back to a theory put forward in the second century AD by Ptolemy of Alexandria, the last great astronomer of ancient times. According to Ptolemy's 'apparent distance' explanation, any object seen across filled space, such as the Moon when near the horizon, will seem to be further away than an object which is seen across empty space, such as the Moon when high up. If the images of the two objects are really equal in size, then the one which seems the further away will give the impression of being larger. I once tested this during a *Sky at Night* television programme by comparing the real Moon with its image seen in a mirror; the mirror could be rotated, and so the altitude of the artificial moon could be altered. I found that the illusion was still evident.

The Moon shows considerable detail when seen with the naked eye. The most obvious features are the large grey plains which are still known as seas even though there is no water in them. In fact there has never been any water on the Moon, and neither is there any atmosphere today. The Moon is only $\frac{1}{81}$ as massive as the Earth, and its gravitational pull is weak; on the lunar surface an astronaut weighs only one-sixth as much as he does at home, and any atmosphere the Moon may once have had has leaked away into space. There are no clouds, no winds and no 'weather'.

The lunar days and nights are long, because the Moon takes not 24 hours, but 27.3 Earth-days to spin once on its axis. This is, of course, the same time as the Moon takes to complete one journey round the Earth, and the result is that the Moon keeps the same face turned towards us all the time, and the markings keep to fixed positions on the disk (or virtually so; there are slight 'wobblings' which need not concern us at the moment).

Look at the full moon, and you will be able to see the main seas without the slightest difficulty. The map below shows most of the features identifiable

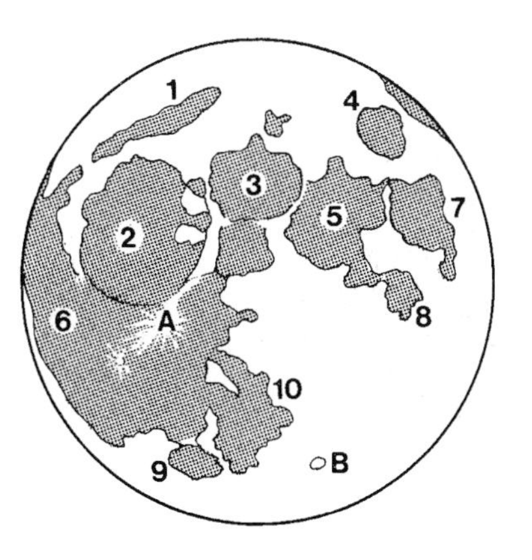

The Moon as seen with the naked eye. Seas: 1 – Mare Frigoris; 2 – Mare Imbrium; 3 – Mare Serenitatis; 4 – Mare Crisium; 5 – Mare Tranquilitatis; 6 – Oceanus Procellarum; 7 – Mare Fœcunditatis; 8 – Mare Nubium. Craters: A – Copernicus; B – Tycho.

with the naked eye; the well-formed mass made up of the Mare Serenitatis (Sea of Serenity), Mare Tranquillitatis (Sea of Tranquillity), Mare Imbrium (Sea of Showers) and Oceanus Procellarum (Ocean of Storms), the small but prominent isolated Mare Crisium (Sea of Crises) and the forked appearance of the Mare Foecunditatis (Sea of Fertility) and Mare Nectaris (Sea of Nectar). There is also the lighter grey strip of the Mare Frigoris (Sea of Cold). The seas are lava-plains, and they must represent tremendous volcanic activity in the remote past, though nothing much has happened there for the last thousand million years at least – probably longer.

The lunar surface is dominated by the walled circular structures known as craters, ranging from huge enclosures well over 100 miles in diameter down to tiny pits so small that from Earth they cannot be seen at all. With the naked eye only a few can be made out. Clavius, nearly 150 miles across, is detectable when it is near the terminator – that is to say the line between the daylit and night hemispheres of the Moon. Two craters, Copernicus and Tycho, each over 50 miles in diameter, show up as bright spots even though they cannot be identified as craters without optical aid.

There are many legends about the Moon. The Man in the Moon is well known, because the arrangement of the bright and darker areas is said to give the impression of a human face, though I admit that I have never been able to recognize it at all clearly. One tale, from Germany, tells that the Old Man was a villager who was caught in the act of stealing his neighbour's cabbages, and was placed in the inhospitable Moon to act as an awful warning to would-be thieves, With any optical assistance the Old Man disappears, drowned in a mass of detail.

During the crescent stage, when the Moon first emerges into the twilight, the most noticeable feature is the Mare Crisium; after full moon, the Mare Crisium is the first of the 'seas' to disappear. Incidentally, it is worth noting that half-moon is only one-ninth as bright as full moon, partly because the intensity of light falls off near the terminator and partly because the mountains and craters cast long shadows. The evening half-moon is considerably brighter than the morning half-moon, because the areas visible after full contain more of the dark plains.

Mountains on the Moon rise to well over 20,000 feet in places. The Apennines, forming part of the boundary of the Mare Imbrium, contain some lofty peaks and are the most spectacular of all the ranges, though not the highest. Indications of them can be seen with the naked eye when the Sun is rising or setting over them. Binoculars, of course, show them well, together with many of the craters, while a telescope brings out so much detail that to map the surface at one sitting would be an impossible task.

By terrestrial standards, the lunar craters are very old; most of them date back for well over a thousand million years. They were formed during a 'cosmical bombardment', when meteoritic bodies rained down upon the lunar surface. At least we now know that the surface layers are firm enough to bear the weight of a spacecraft.

It seems a long time now since Neil Armstrong, from Apollo 11, made his 'one small step' on to the barren landscape of the Sea of Tranquillity. Since then there have been five more successful expeditions to the Moon, the last of which was in December 1972, while many unmanned spacecraft have also been there. We have been able to analyse the lunar rocks, which are mainly volcanic, and also to map the far side of the Moon, which we can never see from Earth. As expected, the regions on the far side are just as barren and cratered as the regions we have always known, though they contain fewer of the large grey plains.

Eclipses of the Moon are quite different from those of the Sun. They occur when the Moon passes into the Earth's shadow, so that all direct sunlight is cut off.

The Earth casts a long shadow in space. The main cone, or umbra, is about 860,000 miles long; this is more than three times the distance between the Moon and the Earth, so that at the Moon's mean distance (239,000 miles) the cone has a diameter of about 5700 miles. When the Moon passes through the centre of the umbra, it may be eclipsed for as long as one and three-quarter hours. And since the Sun is a disk, not a point source of light, the umbral cone is flanked to either side by an area of partial shadow, the penumbra, as shown in the diagram.

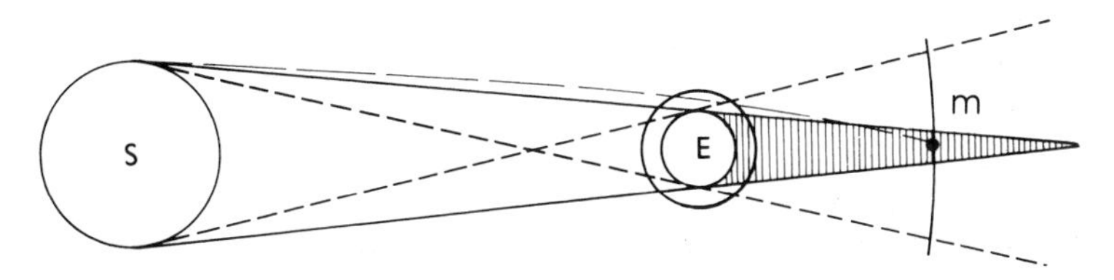

Theory of a lunar eclipse. S – Sun; E – Earth; m – the position of the Moon at mid-totality. Not to scale.

The Moon does not usually disappear, even at mid-totality, because the shell of atmosphere surrounding the Earth acts rather in the manner of a lens, and bends some of the sunlight on to the lunar surface. At most eclipses the Moon merely turns a dim, rather coppery colour until it emerges from the shadow. Not all eclipses of the Moon are total; some are partial, and there are also penumbral eclipses, when the Moon misses the main cone altogether. These are hard to detect with the naked eye, and nothing can be made out apart from a very slight dimming of the surface, but they are worth looking at.

Observers with telescopes often record lovely colours during full eclipses; reds, greens and blues may be seen, depending upon the state of the Earth's air, since any sunlight reaching the eclipsed Moon has to pass through our atmosphere en route. Unfortunately these colours are not well seen with the naked eye, though they are sometimes detectable.

The main interest for the naked-eye observer concerns the brightness of the eclipsed Moon. Sometimes it is very obvious throughout, but occasionally the Moon vanishes completely. The eclipses of 1963 and 1964, for example, were very dark, because a violent volcanic outburst from Mount Agung, in the East Indies, had sent a vast quantity of dust and ash high into the atmosphere. On the other hand, it is said that the eclipse of 1848 was so bright that casual observers did not notice it at all.

When a lunar eclipse occurs, it is visible from any point on the Earth from which the Moon is above the horizon. Eclipses between 2000 and 2008 are as follows:

2000	21 Jan	Mid-eclipse	04.45 G.M.T.	Total for 1 h 16m
2000	16 Jul		13.57	Total for 1 h 0m
2001	9 Jan		20.22	Total for 30m
2001	5 Jul		14.57	Partial: 49% eclipsed
2003	16 May		03.41	Total for 25m
2003	9 Nov		01.20	Total for 11m
2004	4 May		20.32	Total for 38m
2004	28 Oct		03.05	Total for 40m
2005	27 Oct		12.04	Partial: 6% eclipsed
2006	7 Sept		18.52	Partial: 18% eclipsed
2007	3 Mar		23.22	Total for 37m
2008	21 Feb		03.27	Total for 24m
2008	16 Aug		21.11	Partial: 81% eclipsed

While dealing with what may be called celestial hide-and-seek, I must say something about occultations. As the Moon moves across the sky, it must sometimes pass in front of stars and hide, or occult, them. There is no atmosphere around the Moon's limb, and so the star shines steadily up to the moment when it is occulted; as the Moon sweeps in front of it, the star snaps out as abruptly as a candle-flame in the wind. When the occultation takes place at the dark limb of the Moon, it is really spectacular. One moment the star is there; the next, it has gone.

Occultations are common, but not as common as most people probably imagine, because the Moon is so deceptively small in the sky. The only first-magnitude stars close enough to the ecliptic to be occulted are Antares, Aldebaran, Pollux, Spica and Regulus. I very much doubt whether the naked-eye observer would be able to follow even Aldebaran, the brightest of them, right up to the moment of occultation except under ideal conditions. There is, however, one optical illusion which is worth noting. As the Moon creeps closer and closer to the star, the unwary watcher may think that the star has moved 'between the horns', so to speak, as in the Turkish flag. Binoculars bring this out well unless the dark side of the Moon is earthlit, but I am afraid that it is beyond the range of the naked-eye observer.

Planets are occulted occasionally, and with Venus, Jupiter and – when at its brightest – Mars, the naked-eye observer can follow the complete phenomenon. However, it does not happen often. Over the past fifty years I have seen only four planetary occultations – two of Mars, one of Venus and one of Saturn. Occultations of stars by bright planets are less frequent still. I was lucky once, in 1959, when Venus passed in front of Regulus, but I was using a powerful telescope, and nothing of the sort will happen again in my lifetime. Nature can provide some remarkable spectacles, but one has to be prepared to wait!

35

The planets

Amateur astronomers as a class are particularly interested in the planets. This is natural enough, because they show definite disks, and some of them are rich in detail provided that one has an adequate telescope. Unfortunately the naked-eye observer can do very little except learn how to recognize the planets and follow their movements against the starry background, and for this reason I do not propose to say a great deal about them here.

Any rough map of the Solar System is enough to show that it is divided into two parts. There are four small inner planets: Mercury, Venus, the Earth and Mars. Beyond Mars there is a wide gap, in which move thousands of dwarf worlds known as the minor planets or asteroids. Outside the asteroid belt come the four giants: Jupiter, Saturn, Uranus and Neptune. The planetary family is completed by Pluto, a peculiar little world which is apparently in a class of its own.

The five naked-eye planets – Mercury, Venus, Mars, Jupiter and Saturn – have been known since very early times. Of the rest, Uranus was discovered in 1781; Ceres, first of the asteroids, in 1801; Neptune in 1846; and Pluto in 1930.

The so-called inferior planets, Mercury and Venus, are closer to the Sun than we are, and have their own way of behaving. Obviously they always seem to lie in the same part of the sky as the Sun; the elongation – that is to say, the angular distance between the Sun and the planet – can never be more than 47 degrees for Venus and is always less than 30 degrees for Mercury, so

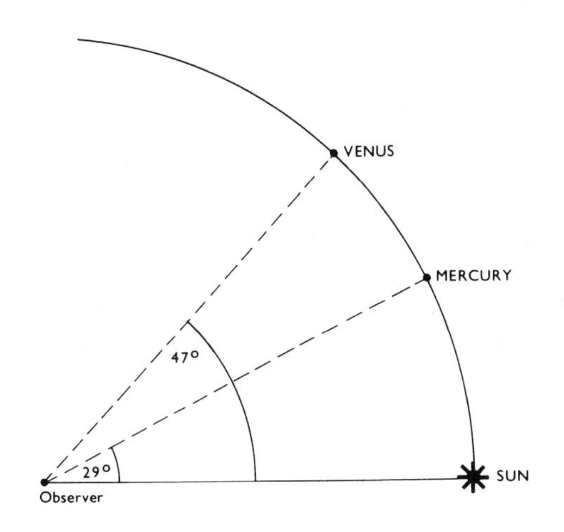

Diagram to show that Mercury and Venus always appear in much the same general direction as the Sun.

that the best views with the naked eye are obtained in the western sky after sunset or in the eastern sky before dawn. Mercury and Venus can never be seen throughout the night.

MERCURY is never particularly easy to see with the naked eye, and under the circumstances it is surprising that the ancient stargazers should have known so much about its movements, though admittedly the Mediterranean skies are relatively clear and dark. With its diameter of only 3030 miles, Mercury is not a great deal larger than the Moon, though it is much more massive. It never comes much within 50 million miles of the Earth, and before the Space Age we knew practically nothing about its surface details.

The Mercurian 'year' is 88 days, so that the planet is a rapid mover (which is why it was named after the Messenger of the Gods). Its nearness to the Sun is a serious handicap, and I very much doubt whether anyone who stays in a town or even a village will be able to find it. From my home in England, on the end of Selsey Bill, Sussex, where I am surrounded by sea on three sides and there is little in the way of artificial lighting, I can see Mercury without optical aid on about twenty-five evenings or mornings each year, though I remember that in 1977, when the weather was generally poor, I saw it only twice.

The most favourable dates for locating Mercury are given in yearly almanacs. Generally speaking, the planet is on view for at least a week to either side of the date of greatest elongation, provided that conditions are favourable. In the northern hemisphere the best opportunities are in the spring, when Mercury is in the evening sky, and autumn, when it is a morning object. (This applies to the northern hemisphere. The same is true of the south, bearing in mind that Britain's and the United States' spring is Australia's and South America's autumn, and vice versa.)

Elongations for the 2000–2005 period are:

Western (morning object)

2000 28 March, 27 July, 15 November
2001 11 March, 9 July, 29 October
2002 21 February, 21 June, 13 October
2003 4 February, 3 June, 27 September
2004 17 January, 14 May, 9 September, 29 December
2005 26 April, 23 August, 12 December

Eastern (evening object)

2000 15 February, 9 June, 6 October
2001 28 January, 22 May, 18 September
2002 11 January, 4 May, 1 September, 26 December
2003 16 April, 14 August, 9 December
2004 29 March, 27 July, 21 November
2005 12 March, 9 July, 3 November

The best way to find Mercury is to sweep with binoculars when the Sun is just below the horizon. Once you have located the planet you will probably be able to see it with the naked eye, and you may even wonder how you could have overlooked it. Actually, Mercury is often brighter than Sirius, but it does not seem so, because it can never be seen against a dark background.

In 1974 the spacecraft Mariner 10 flew past Mercury, and sent back the first pictures from close range. Mercury turned out to be mountainous and cratered; there is almost no atmosphere, and the temperature conditions are most unpleasant. Mercury takes 58½ Earth-days to spin once on its axis, leading to a very peculiar sort of calendar.

Mercury shows phases like those of the Moon, but they are quite beyond naked-eye range. Sometimes, too, Mercury will pass in front of the Sun, in transit, and telescopes then show it as an inky-black disk, but it is not visible without optical aid. The next transit will be on 7 May 2003.

VENUS, with a diameter of 7520 miles, is nearly as large and massive as the Earth. It is as different from Mercury as it could possibly be. Instead of having no atmosphere, it has so much that the true surface is permanently hidden. The weather on Venus is always cloudy.

Moving in an almost circular orbit 67 million miles from the Sun, Venus takes 224 days to complete one revolution. People with access to good binoculars will be able to see the phases without difficulty. When at position 1 in the diagram below, at what is termed inferior conjunction, it lies between the Sun and the Earth; its dark side is turned toward us, so that Venus is 'new' and cannot be seen except when the alignment is perfect, so that there is a transit. At such times Venus shows up as a black disk against the Sun, so conspicuous that it is easily seen with the naked eye. I cannot give a personal account, because I have never seen a transit of Venus – and I am fairly sure that the same is true of everyone now alive, because the last transit took place as long ago as 1882. The next is not due until 2004.

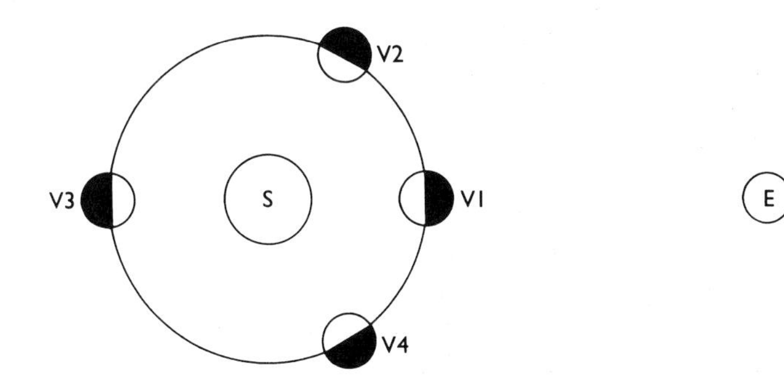

From position 1, Venus moves along and starts to show us a little of its sunlit side. (To make matters easier, I am assuming that the Earth is motionless; of course this is not true, but for the moment we can ignore our own motion round the Sun.) Venus appears in the evening sky as a crescent; the phase increases until elongation (2), when we see the half disk. The phase continues to grow until Venus is full, at superior conjunction (3), but as it is then on the far side of the Sun it is out of view. Venus then reappears in the morning sky, first as a three-quarter or gibbous shape, then half-phase again at western elongation (4), after which it narrows to a crescent before returning to new at the next inferior conjunction.

This is not really a digression, because there is little doubt that during the crescent stage there are a few very keen-sighted people who can see the phase with the naked eye. My sight is not good enough for this, but I know two observers who can see the crescent at any time when conditions are favourable. Try it by all means – but do not be disappointed if you fail.

Elongations for 2000–2005 are as follows:

Western (morning object)
2001 17 January
2002 22 August
2004 29 March
2005 3 November

Eastern (evening object)
2001 8 June
2003 11 January
2004 17 August

Venus is much brighter than any other star or planet; its magnitude may reach −4.4, so that it is then almost three magnitudes brighter than Sirius. It is detectable with the naked eye for some time after sunset or before sunrise when well placed, and those with exceptional sight can locate it even in broad daylight. I am reminded of a quote from the nineteenth-century French astronomer François Arago, concerning no less a person than Napoleon. When the Emperor was attending an official fête in Luxemburg, he was 'very much surprised at seeing the multitude pay more attention to the regions of the heavens above the palace than to his person or the brilliant staff accompanying him. He inquired the cause, and learned that these curious persons were observing with astonishment, although it was noon, a star, which they supposed to be that of the Conqueror of Italy; an allusion to which the illustrious general did not seem indifferent when he himself with his piercing eyes observed the radiant body.'

Venus can be so bright that at its best, shining down from a darkened sky several hours after sunset or before sunrise, it can cast shadows. It is interesting to experiment with a white screen; the shadows are very sharp, since Venus appears almost as a point source of light. And when the planet is seen shining above a sea or lake, the 'light trail' across the water is really beautiful. I remember seeing this excellently in the late summer of 1964, from Loch Ness in Scotland; the trail stretched right across the loch, though to my regret it failed to show any trace of the famous monster.

Though named after the Goddess of Beauty, Venus has turned out to be a very hostile world – possibly the most unfriendly in the whole of the Solar System. The Russians have soft-landed four probes there, and have obtained pictures direct from the surface, while both American and Russian vehicles have been put into orbit round the planet and have mapped the surface by radar. Venus is a world of plains, two major uplands, and what are certainly volcanoes, probably active. The surface temperature is not far short of 1000 degrees Fahrenheit (hotter even than Mercury); the atmospheric pressure is over 90 times that of our air at sea-level; the atmosphere itself is made up chiefly of the heavy gas carbon dioxide, which acts as a greenhouse and shuts in the Sun's heat; and those lovely, shining clouds contain a great deal of corrosive sulphuric acid. I would not recommend a trip to Venus, and it seems unlikely that anyone will go there, at least in the foreseeable future. Like Mercury, it has no satellite.

The other planets lie beyond the Earth's orbit in the Solar System, so that they do not show phases from new to full. (This is obvious enough. To appear as a crescent, a planet must be more or less between the Earth and the Sun; only Mercury and Venus can be so placed.) First on the list is MARS, which may approach us to within a distance of 35 million miles. It never comes as close as Venus, and it is much the smaller of the two, with a diameter of only 4200 miles. Neither is it so good a reflector of sunlight. To make up for this, Mars is observable high in the sky against a dark background. At its best it may outshine every other body apart from the Sun, the Moon and Venus.

Mars takes 687 days to go once round the Sun, at a mean distance of 141,500,000 miles. It is most favourably placed when at opposition – that is to say, exactly opposite to the Sun in the sky. This does not happen every year, as the diagram (overleaf) will show. Begin with the Earth at E1 and Mars at M1; the planet is then at opposition. A year later the Earth has completed its circuit, and has arrived back at E1; but Mars, moving more slowly in a larger orbit, has only moved as far as M2. The Earth has to catch up with Mars, so to speak, and does so when the Earth has reached E2 and Mars has moved on to M3. On average, the interval between one opposition and the next (the 'synodic period') is 780 days. Thus there was an opposition in 1999, and there will be one in 2001, but not in 1998 or 2000.

To complicate matters, the orbit of Mars is much less circular than ours, and whenever opposition takes place when the planet is near aphelion (its greatest distance from the Sun), as in 1980, it does not come closer than

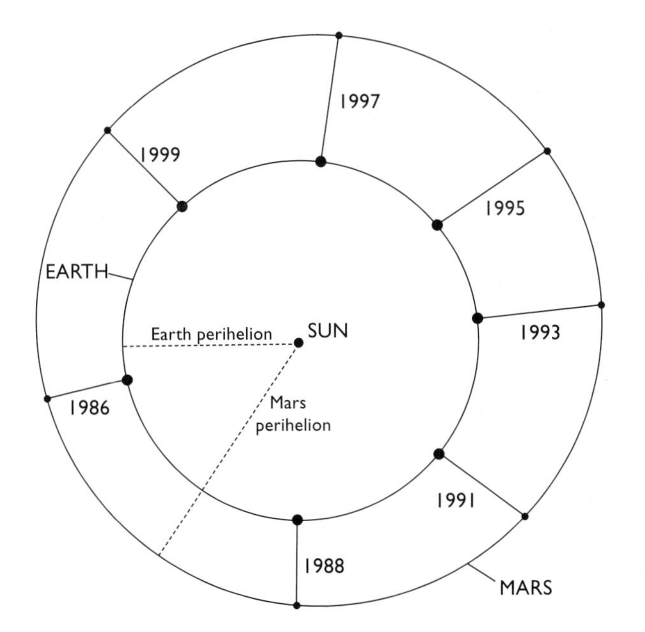

Oppositions of Mars, 1986–1999.

62 million miles. It was then of about magnitude −1, brighter than Canopus but not so bright as Sirius. At oppositions near perihelion (closest approach to the Sun), Mars is much more brilliant. The next really good opposition will be that of 2003.

The opposition dates up to 2005 are as follows – though remember that the exact date of opposition is not important, as the planet is well seen for several weeks to either side of opposition itself:

2001 13 June Maximum magnitude −2.1; Mars will be in Sagittarius.
2003 28 August Maximum magnitude −2.7; Mars will be in Capricornus.
2005 7 November Maximum magnitude −2.1; Mars will be in Aries.

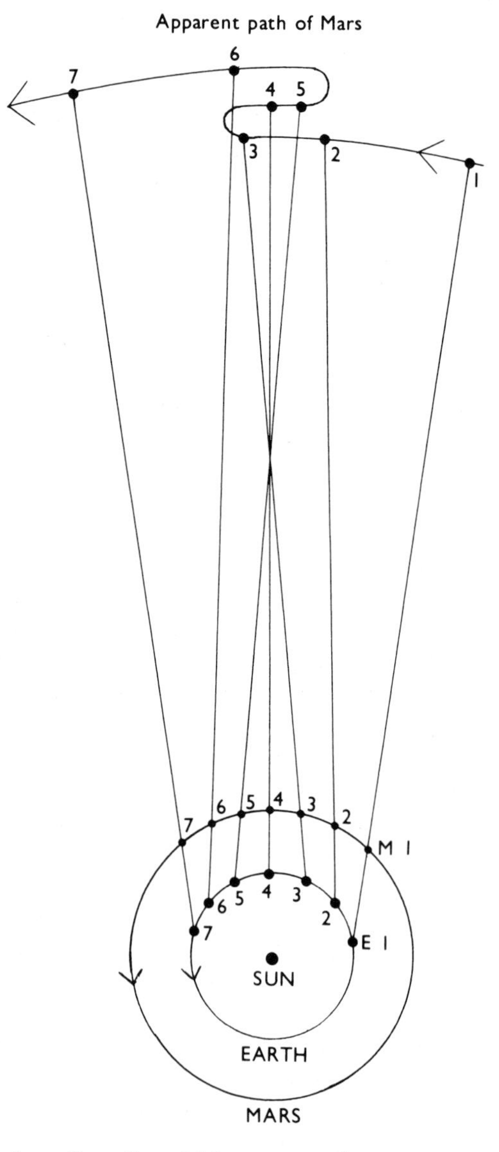

Retrograde or 'backward' motion of Mars among the stars.

One problem about identifying Mars is that it is not always brilliant, and when at its greatest distance from us it is little brighter than the Pole Star, though it is always very red.

If you check the changing position of Mars against the starry background, identifying a few nearby stars and making charts every few nights, you will notice something strange. Although the usual trend is from west to east, there are times when Mars appears to stop and then move in a retrograde or east–west direction before stopping again and resuming its eastward motion.

The diagram shows what really happens. The Earth's speed round the Sun is greater than that of Mars: about 18½ miles per second as against 15. When the Earth is at E1 and Mars at M1, Mars will appear in the sky in position 1, and so on. There will be a period, when the Earth is 'passing' Mars, when the apparent motion of the planet is reversed, and Mars describes a slow loop in the sky.

Mars has always been of special interest because until quite recently (that is to say, before 1965) it was thought that life might exist there. Telescopes show dark areas and broad ochre 'deserts', together with white polar ice-caps which wax and wane with the Martian seasons. But in 1965 Mariner 4, the first successful Mars probe, flew past the planet and sent back data which caused a rapid change in outlook. Since then there have been many other missions, crowned by the two Vikings, which soft-landed in 1976 and transmitted impressive views of a reddish, rock-strewn landscape under a pink sky. Since then further probes have been sent there: in 1997 the Pathfinder vehicle carried a small rover, Sojourner, which was able to move around and analyse the Martian rocks. There are mountains and craters on Mars, together with huge volcanoes, one of which rises to a height of 15 miles above the outer country; but there are no seas or lakes, and the atmosphere is very thin. No signs of life have been found. It is too early to claim that Mars is sterile, but most people believe so. In a way, I suppose that Mars has been a disappointment; but it will be reached by men in the foreseeable future, if all goes well, and it is much less hostile than either Mercury or Venus. Mars has two tiny satellites, Phobos and Deimos, neither of which is as much as 30 miles in diameter, and which are far beyond the range of any but fairly large telescopes.

We can pass quickly over the minor planets or asteroids, most of which keep to the gap between the orbits of Mars and Jupiter. Ceres, 584 miles in diameter, is the largest of them, but is below naked-eye visibility. The only member of the swarm ever visible without optical aid is Vesta, which is smaller than Ceres but closer and more reflective. It can reach magnitude 5.7, so that it is not easy to locate without binoculars or a telescope, and of course it appears only as a speck of light. Its mean distance from the Sun is 219 million miles.

Well beyond the main asteroid belt we come to JUPITER, the giant of the Sun's family. It is 88,700 miles in diameter as measured through its equator (somewhat less as measured through the poles, because its globe is decidedly flattened), and it has a mass over 300 times that of the Earth. It moves round the Sun in a period of nearly 12 years, at a mean distance of 483 million miles.

The synodic period of Jupiter is 399 days, so that it is well seen for several months in every year. It is easy to recognize, because it is so bright; it is outshone only by Venus and, on rare occasions, by Mars. Oppositions in the 2000–2005 period are:

2000	28 November	Jupiter will be in Taurus
2002	1 January	Jupiter will be in Gemini
2003	2 February	Jupiter will be in Cancer
2004	4 March	Jupiter will be in Leo
2005	3 April	Jupiter will be in Virgo

The year 2001 is missed out, and the next opposition will be that of 1 January 2002. The opposition magnitude does not vary much from its mean value of −2.3.

Several spacecraft have flown past Jupiter – notably Pioneers 10 and 11 in 1973 and 1974 respectively, the two Voyagers in 1979 and Galileo, which reached the planet in 1995. Jupiter has proved to be a remarkable world. It is not solid and rocky; the surface is gaseous, showing belts and spots, and is in a state of constant turmoil. One notable feature is the Great Red Spot, a huge oval with a surface area larger than that of the Earth, now known to be a kind of whirling storm. Seen through a telescope Jupiter is magnificent, but it is also menacing inasmuch as it is surrounded by a zone of lethal radiation which would quickly kill any astronaut careless enough to venture too close.

So far as the naked-eye observer is concerned, the main interest is in the four large satellites, which were discovered as long ago as 1610 and which are easy to see with any optical aid. They are named Io, Europa, Ganymede and Callisto. Io is slightly larger than our Moon, Europa slightly smaller, and Ganymede and Callisto much larger; indeed, Ganymede is larger than Mercury, though less massive.

The Pioneer and Voyager pictures were astonishing. Ganymede and Callisto have icy, cratered surfaces; Europa, also icy, is as smooth as a billiard-ball, while Io is yellowish-orange, with a sulphur coating and violently active volcanoes. Since Io also moves in the thick of Jupiter's radiation zone, it is a world to be viewed from a respectful distance. The revolution periods round Jupiter range from 1 day 18½ hours for Io up to 16 days 16½ hours for Callisto.

Can the satellites be seen with the naked eye? Ganymede at least would be easy were it not so overpowered by Jupiter, and there is evidence that it was glimpsed by a Chinese observer named Gan De in the year 364 B.C. The best chance is to choose a moment when two of the satellites are close together, so that they show up as one object. Try by all means, but you will need a perfectly clear, dark sky with Jupiter high above the horizon.

SATURN was the outermost planet known in ancient times. It is smaller than Jupiter, though its equatorial diameter is 75,000 miles; it moves round the Sun at a mean distance of 886 million miles in a period of 29½ years. The synodic period is 378 days, so that Saturn comes to opposition about a fortnight later each year; oppositions occur on 19 November 2000, 3 December 2001, 17 December 2002, 31 December 2003 and 13 January 2005. Because Saturn moves slowly, it stays in the same constellation for some time.

Saturn is the gem of the sky. It is surrounded by a magnificent system of rings, made up of icy particles whirling round the planet in the manner of dwarf moons. The ring-system measures 170,000 miles from end to end but is less than a mile thick, so that when the rings are edgewise-on to us, as last happened in 1995, they almost disappear even with powerful telescopes. When the rings are 'open', as in the mid-1980s, they are superb even with a small instrument. The Voyager probes, which by-passed Saturn in 1980 and 1981 respectively after having left Jupiter far behind, have shown that the rings are incredibly complex; there are thousands of narrow ringlets and minor divisions, and it has been said that the ring system contains more 'grooves' than a gramophone record. The exact reason for this is still uncertain, but it must have something to do with the satellites, of which eighteen are now known – though even Titan, much the largest of them (over 3200 miles in diameter), is well below naked-eye visibility. Titan, incidentally, has a thick atmosphere made up chiefly of nitrogen, and even the Voyagers could show no more than the top of a layer of orange 'smog'.

There is not the slightest chance of seeing Saturn's rings with the naked eye, and even strong binoculars will do no more than show that there is something unusual about the shape of the planet. But if you ever have the chance to turn a telescope towards it, do not miss the experience. For sheer beauty Saturn is, in my opinion, unrivalled.

Until 1781 it was thought that the Solar System must be complete. It was a major surprise when William Herschel, then an obscure amateur astronomer, discovered an object which proved to be a new planet, now called URANUS.

Herschel did not at once realize what he had found. He mistook it for a comet, and not until its orbit was worked out did astronomers learn that it was something much more important. Uranus is an icy giant; its diameter is 30,000 miles, so that it is by no means the equal of Jupiter or Saturn, but it could still hold fifty globes the volume of the Earth.

Though Uranus is so remote (1,783 million miles from the Sun at its mean distance), it is distinctly visible with the naked eye when best placed. The mean opposition magnitude is 5.7, and keen-sighted people will be able to see it easily once they know where it is. It is a slow mover.

NEPTUNE, slightly smaller but more massive than Uranus, is much too faint to be seen with the naked eye, though binoculars will show it. It need not concern us here – and neither need PLUTO, which is smaller than the Moon and is accompanied by a satellite, Charon, one-third the size of Pluto itself. To see Pluto at all you need a powerful telescope. Its orbit is markedly elliptical, and between 1979 and 1999 it was closer in than Neptune, thereby temporarily forfeiting its title of 'the outermost planet'. (There is no fear of a collision on the line, because Pluto's orbit is tilted at the relatively sharp angle of 17 degrees.) Actually, the status of Pluto is very much in doubt. It simply does not fit into the general pattern, and it may not be worthy of true planetary rank.

There are rare occasions when two or more planets lie close together in the sky. These conjunctions are not in the least important, but they are interesting to watch. Even occultations of one planet by another are not unknown:

thus on 3 October 1590 Venus passed in front of Mars, and since the whole phenomenon was watched by Michael Mastlin, Professor of Mathematics at the university of Heidelberg, it must have been visible with the naked eye, since telescopes did not come upon the scene until many years later. On 21 July 1859 Venus and Jupiter passed so near each other that they could not be seen separately with the naked eye, and there are various other records of close approaches – due, of course, to nothing more fundamental than line-of-sight effects.

Planetary groupings also occur sometimes. Thus in February 1962 there was a gathering of planets in Capricornus, including Mercury, Venus, Mars, Jupiter and Saturn. It was invisible with the naked eye, or indeed with a telescope, because the Sun was not far off and all the planets passed below the horizon very soon after sunset, but astrologers and other eccentrics made the most of it. In India, where astrology is still a force to be reckoned with, there was widespread panic and the end of the world was confidently expected, which only goes to show that we cannot afford to laugh at our ancestors who believed the Earth to be flat. Another conjunction of several planets occurred in May 2000, but was not visible with the naked eye, because it was too close to the Sun.

It is true that no detailed observations of the planets can be made without a telescope, but this is no reason for ignoring them. The observer who tracks down Mercury in the evening twilight, who tries to see the crescent of Venus, checks the magnitude and movement of Mars, identifies Jupiter and Saturn, and does his best to make out the dim speck of Uranus may not be contributing to scientific knowledge, but he is at least teaching himself – and enjoying himself at the same time.

36

Space nomads

Comets and meteors have been termed the nomads of the Sun's family. They belong to the Solar System, but they are very definitely the junior members, and many of them are short-lived on the cosmical time-scale. Meteors, indeed, cannot be seen at all until the last few seconds of their lives, when they burn away in the Earth's atmosphere.

Most people picture a comet as being a brilliant object, with a long tail stretching across the sky. This is a fair description of a 'Great Comet', but the average comet is unspectacular, often tailless and too faint to be seen with the naked eye.

It is often thought, too, that a comet will race across the sky very rapidly. This is quite wrong. Comets are millions of miles away, and have to be watched for hours before it is possible to notice any obvious shift against the starry background. If the object is moving perceptibly, it must be either a meteor or an artificial satellite (unless, of course, it is something much more mundane, such as an aircraft or a high-altitude weather balloon catching the sunlight).

Comets are not solid bodies, and they are of very low mass. A large comet consists of a central nucleus, surrounded by a head or 'coma' made up of gases and what may be termed dust, There will be at least one tail, and probably two. Not that all comets have tails; many of the smaller ones do not, and look like tiny, fuzzy patches. Charles Messier catalogued star-clusters and nebulae to avoid confusing them with comets; certainly a comet may look exactly like a nebula when seen through a small telescope, of the size which Messier used.

Comets are bona-fide members of the Solar System. They move round the Sun, but in most cases their orbits are very elliptical. Indeed, all the bright comets (apart from Halley's, of which more anon) take so long to complete a single journey round the Sun that their periods may amount to thousands or even millions of years, and obviously we cannot predict them. They depend upon sunlight for their luminosity, and can be seen only when they are reasonably close in. Few comets can be tracked out beyond the orbit of Jupiter, and most comets are lost when still between the orbits of Saturn and Uranus.

It is thought that long-period comets come from a 'cloud' of icy bodies orbiting the Sun at a distance of about a light-year; this was suggested by the Dutch astronomer Jan Oort, and is known as the Oort Cloud. If one of them is perturbed for any reason, it may leave the cloud and swing inward towards the Sun. One of two things may then happen. The comet may whirl round the Sun with a very small perihelion distance and then return to the cloud, not to make another inward dash for a very long time. Alternatively, it may be 'captured' by a planet (usually Jupiter) and forced into a much smaller orbit, which brings it back to perihelion every few years. Many of these short-period comets are known but most are thought to come from a much closer swarm, the Kuiper Belt, not very far beyond the orbit of Neptune. The comet with the shortest period of all is Encke's, which takes a mere 3.3 years to complete one circuit; its distance from the Sun ranges between 31 million miles (slightly inside the orbit of Mercury) out to 380 million miles (well beyond the main asteroid belt). Unfortunately all the comets with periods of less than a few centuries, again apart from Halley's, are too dim to be seen without optical aid, though just occasionally one of them may reach the fringe of naked-eye visibility, as D'Arrest's periodical comet did in 1976.

The really brilliant comets must be truly magnificent. I have yet to see one, because the last, the so-called Daylight Comet, appeared as long ago as 1910. Earlier, in the nineteenth century, there had been bright comets quite often – in 1811, 1843, 1858, 1861 and 1882, for example. The coma of the 1811 comet was larger than the Sun, and the tail of the great comet of 1843 stretched out for a distance greater than that between the Sun and Mars. Yet even these comets were flimsy by planetary standards, and could not possibly pull any of the planets out of position.

In most cases, comets are named after their discoverer or co-discoverers; thus Faye's Comet was discovered by the French astronomer Hervé Faye, D'Arrest's Comet by Heinrich D'Arrest, and so on. Less frequently the comet is named after the mathematician who first computed the orbit; Encke's Comet is an example of this. Some of the names are tongue-twisting. Thus we have Comets Schwassmann–Wachmann, Churyumov–Gerasimenko and Honda–Mrkós–Padjusaková.

Two bright comets appeared near the close of the twentieth century. Comet Hyakutake, of 1996, was a bright naked-eye object for a few nights, and had a long tail; it came close to us – no more than 10,000,000 miles away at its nearest – but will not return for 15,000 years. It was outmatched by Comet Hale–Bopp of 1997, which was a splendid sight and remained visible with the naked eye for many weeks; it had two main tails, and its nucleus became brighter than most stars. Had it come as close to us as Comet Hyakutake had done, it would have remained visible in broad daylight. It has a long period, and will not return for over 2000 years.

When a comet is a long way from the Sun, it will have no tail. As it draws inward, the nucleus, made up chiefly of ice with embedded 'dust', begins to evaporate; the nucleus itself is effectively hidden, and material is driven out to make a tail or tails. If the tail is of gas it will be straight; if of dust, it will be curved, because the dust will tend to lag behind. Donati's Comet of 1858 was the supreme example of a brilliant comet which looked like a scimitar, with both the gas and dust tails very striking. (Unfortunately it came too early to be photographed, but at least we have some beautiful drawings of it. It is said to have been the loveliest comet ever observed.)

Tails have one peculiarity; they always point more or less away from the Sun, because the tiny particles in them are repelled by what is called the solar wind – a stream of atomic particles moving outward from the Sun constantly

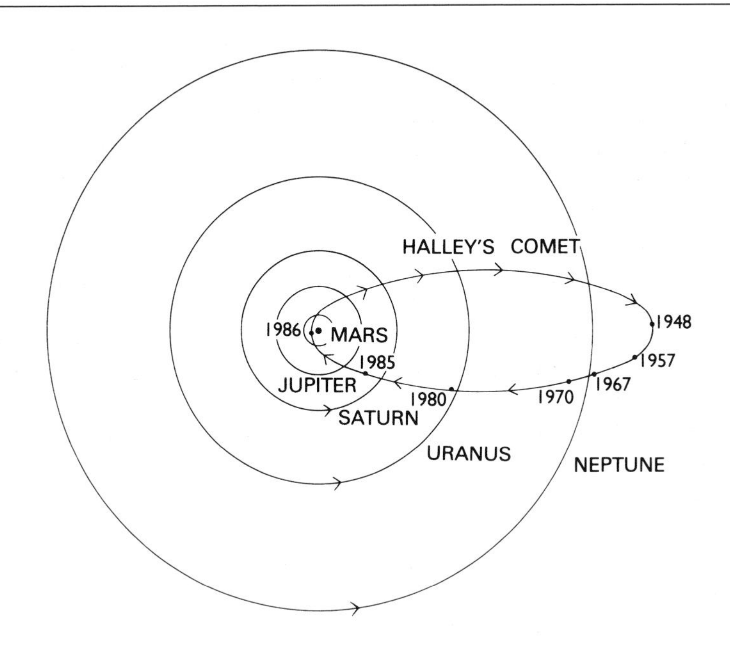

Orbit of Halley's Comet. Aphelion – 1984; perihelion – 1986.

in all directions – and by the pressure of sunlight. Therefore, as the comet rounds the Sun at perihelion the tail swings, and during its outward journey the comet moves tail-first. Occasionally, as with the naked-eye Arend–Roland Comet of 1957, there may be a sunward-pointing 'spike', but this is not a true tail; it is material in the comet's orbit, so placed that it is lit up by the Sun.

Of course the most famous comet of all time is Halley's which has a long history. Records of it go back long before the time of Christ and since 83 BC it has been seen at every return. The period is 76 years (give or take a year or two either way; because of perturbations by the planets, the period is not absolutely constant). It appeared in 1682, and was observed by Edmond Halley, who later became Britain's Astronomer Royal. Later, Halley worked out the orbit, and decided that it was the same comet as those which had been observed in 1607 and 1531. Accordingly, he predicted the next return for 1758. By then Halley was dead; but the comet was duly recovered on Christmas Night, 1758, by a German amateur named Palitzsch, and it reached perihelion in 1759. It came back on schedule in 1835, and again in 1910. It was followed until 1911, after which it was lost to view until astronomers at Palomar Observatory, in California, picked it up once more in October 1982. It reached perihelion in February 1986, though it was not very well placed and never became as conspicuous as it had been in 1910.

Several spacecraft were sent to Halley's Comet in March 1986, and one of these, the European-built Giotto, penetrated the comet's head and sent back close-range images of the nucleus, which proved to be peanut-shaped, with a longest diameter of less than ten miles. It had a dark crust overlying the icy core, and dust-jets were recorded in restricted areas on the sunward side. Halley's Comet will be back in 2061, but will again be rather badly placed for observation.

The naked-eye comet observer can make himself useful in several ways. First, visual estimates of a comet's brightness are always valuable; it is never easy to compare a blurred patch with a sharp point such as a star, but at least some sort of estimate is possible. For example, I well remember Bennett's Comet of 1970, one of the best of recent years; at its brightest, the nucleus was about equal to Altair. Also, there is always the chance of discovering a comet. The Daylight Comet of 1910 (not identical with Halley's) was found by chance by some miners in South Africa, who happened to look upward and realized that something very unusual had been added to the sky.

Remember to make careful checks. During the past couple of years I have had letters from five observers who claimed to have found a comet; alas, in each case what they had done was to rediscover the Andromeda Nebula.

Finally, drawings of the tail or tails of bright comets are valuable, because the tails can change in structure with remarkable speed. This of course is possible only with a fairly bright naked-eye comet, such as Hale–Bopp of 1997.

The age-old terror of comets is quite unfounded. Even a direct hit would do no more than local damage, and there is indeed a strong chance that a small comet, or part of one, crashed in Siberia in 1908, blowing pine-trees flat in all directions. Though the gases in a comet's tail would be poisonous if concentrated, they are much too rarefied to do any damage at all. On at least two occasions during the past hundred and fifty years the Earth has passed through the tail of a comet without any unusual phenomena being noticed.

Meteors are cometary débris, and are 'left behind' as a comet moves along its orbit. The average meteor is smaller than a pin's head, and is very fragile. When it enters the upper air, moving at anything up to 45 miles per second, it has to force its way through the air-particles; this sets up heat by friction, and the meteor is destroyed in the streak of radiance which we call a shooting-star. What we are seeing is not the tiny meteor itself, but the luminous effects in the atmosphere produced by the meteor's headlong plunge to destruction.

Most meteors become luminous at about 120 miles above ground level, and are burned away by the time they have dropped down to 40 miles. They finish their journey in the form of fine dust.

Meteors tend to travel round the Sun in swarms, each of which is presumably connected with a comet even though in some cases the parent comet is no longer to be seen. Each time the Earth passes through a shoal of meteors, we see a shower of shooting-stars, as happens many times in each year. The most spectacular shower is that of the Perseids, which lasts from about 27 July to 17 August, and is so named because its meteors seem to come from the direction of the constellation Perseus. Other important showers are the Geminids of December and the unreliable but occasionally spectacular Leonids of November.

The meteors in any particular shower seem to emanate from a definite point or radiant. The best way to explain this is to consider the scene when you stand on a bridge overlooking a motorway, and watch how the lanes of the motorway seem to meet at a point in the distance. Obviously the lanes do not really converge; what you see is simply an effect of perspective, and it is the same with shower meteors, which are moving through space in parallel

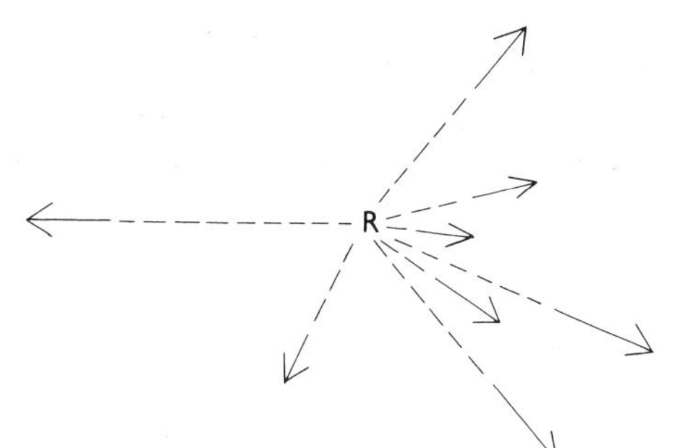

Meteor radiant

paths. There are also sporadic meteors, which do not belong to any known shower, and may flash into view from any direction at any moment.

There are numerous annual showers, though not all of them are rich. Their intensities are measured in terms of what is called the Z.H.R., or Zenithal Hourly Rate, which is simply the average number of meteors which would be seen at peak activity by an observer under ideal conditions, with the radiant of the shower at the zenith or overhead point – conditions which, alas, are virtually never attained.

The main annual showers are:

	Date			Z.H.R.	Notes
	beg.	max.	end.		
Quadrantids	1 Jan	3 Jan	6 Jan	110	Short, sharp maximum
Lyrids	19 Apr	22 Apr	24 Apr	12	Swift meteors
Eta Aquarids	2 May	4 May	7 May	20	Swift; long trails
Delta Aquarids	15 Jul	28 Jul	15 Aug	35	
Perseids	25 Jul	12 Aug	18 Aug	68	Very consistent
Orionids	16 Oct	21 Oct	26 Oct	30	
Taurids	20 Oct	4 Nov	25 Nov	12	Slow meteors
Leonids	15 Nov	17 Nov	19 Nov	var.	Brilliant occasionally
Andromedids	15 Nov	20 Nov	6 Dec	5	Now very feeble
Geminids	7 Dec	14 Dec	15 Dec	58	Rich; fairly swift
Ursids	17Dec	22 Dec	24Dec	12	Radiant in Ursa Minor

The Quadrantids come from that area once separated out as the constellation of Quadrans, the Quadrant. The nearest bright star to the position of the radiant is Beta Boötis. The maximum is very short, lasting for only a few hours, but the meteors can sometimes be very plentiful. The Lyrids are associated with Thatcher's Comet of 1861, the Taurids with Encke's Comet, the Ursids with Tuttle's periodical comet, and the Eta Aquarids and the Orionids with Halley's Comet.

The Perseids – usually very rich – are also associated with a comet, Swift–Tuttle, which was seen in 1862. It has a period of 130 years and so last returned in 1992. The Leonids, associated with Comet Tempel–Tuttle (period

32.9 years), can sometimes produce magnificent displays; this happened in 1799, 1833 and 1866, and records have been traced back as far as the year 933, so that major showers have tended to recur every 33 years. In 1899 and 1933 the Leonids were sparse because their paths had been perturbed by the giant planets, but in 1966 they were back in their full glory, though unfortunately for British observers the maximum lasted for only a few hours and took place during European daylight. For a while on the night of 17 November, observers in Arizona recorded a rate of 60,000 Leonids per hour.

During the 1966 shower I was anxious to try an experiment, so on my television programme *The Sky at Night* I asked for help. We distributed over 10,000 charts, and called upon volunteers to use them to plot the positions of any meteors observed between midnight and dawn on 17 November. We had almost all the 10,000 charts back. Few Leonids were seen, but the experiment was more valuable than it would have been if we had seen the main shower, because it proved that the Leonids are bunched up instead of being spread all around the comet's orbit, as with the Perseids. The whole venture was a decided success, though some of those who took part were disappointed. I still have a letter written to me by a gentleman in Manchester: 'As requested, watched the sky from 12 midnight to 6 a.m. Meteors – from the sky, none. From the wife – plenty!'

The Leonids were fairly rich in 1998, and then, on 18 November 1999, there was a brief but very energetic 'meteor storm'. It was spectacular, though it did not rival the displays of 1866 or 1966.

The Andromedids are different. They are associated with Biela's Comet, which used to move round the Sun in a period of 6¾ years, and was seen regularly. At its return in 1846 it staggered astronomers by breaking in half. The 'twins' came back in 1852, but they have never been seen since. Biela's Comet is dead, but we still see meteors coming from the position of the old comet, though in recent years the shower has become very feeble and may die out completely in the near future.

A Z.H.R. figure can be misleading, since shower radiants almost never lie in the zenith, and a correction has to be made. First, estimate the altitude of the radiant. This will give a correction factor. Multiply your observed hourly rate by the correction factor to give the true Z.H.R.:

Altitude of radiant (degrees)	Correction factor
90	1
66	1.1
52	1.25
42	1.4
35	1.6
27	2
20	2.5
14	3.3
9	5
2.5	10

Suppose that you observe for a full hour, and record 14 meteors – what is the Z.H.R.? Check the altitude of the radiant; let us say that it is 60 degrees. From the table, the correction factor falls between 1 and 1.25; take it as 1.2. Therefore $14 \times 1.2 = 16.8$, which is the true Z.H.R., though in practice it would be rounded off to 17. Of course, it is more complicated if the sky is hazy or moonlit, or not really dark, since faint meteors will be missed and the calculated Z.H.R. will be too low. Beware too of sporadic meteors; sorting them out is largely a matter of common-sense.

The altitude of the radiant changes from hour to hour, so that the correction factor alters too. It is hopeless to expect a high hourly rate when the radiant is low down; some of the meteors will be below the horizon, and others hidden by haze or mist. From Britain, for example, the Leonids are best seen in the early hours of the morning, since in November the radiant does not rise until about midnight.

For once in a way, observers in the northern hemisphere have the best of matters. There are not so many far southern showers, though the Corona Australids of 14–18 March and the Phoenicids of 4–5 December can usually muster a Z.H.R. of about 5.

Meteor observation is one field in which the naked-eye amateur can do really useful work. The method is to plot the paths of the meteors against the stars. Hold up a short stick (a pencil will do quite well) along the line where the meteor has been seen; if you know your way around the sky you can then plot the trail on a chart. Note the exact time, duration of the meteor, magnitude, and any special features such as colour or 'bursting' and flaring. If the same meteor is observed by two observers at different sites, it will appear in different parts of the sky, and its real height and path can then be worked out. The American Meteor Society, as well as the Meteor Section of the British Astronomical Association, has organized many such programmes over the years, and produced results which professional astronomers have found very valuable indeed.

Now and then you may see a fireball, which is simply a rather vague term for a brilliant meteor; I have seen several which have outshone Venus, and two or three which have been even brighter than the Moon. When a fireball is seen, plot the path as accurately as possible, and note down all the relevant details; for instance, there may be a visible trail which persists for several minutes.

Try your hand, too, at meteor photography, by aiming the camera at the meteor radiant and leaving the shutter open. If a meteor is obliging enough to streak across the field of view, you should record it. Generally it is unwise to leave the shutter open for more than half an hour at the most because of dewing, but everything depends upon local conditions.

Meteor watching can be a cold business, so wrap up well and make yourself comfortable in a chair, preferably a deckchair so that you can lean back. In the well-organized programmes of societies such as those mentioned above, it is common practice for four observers to arrange themselves so that each covers one quarter of the sky. But do not attempt too lengthy a spell; once you become really tired you are bound to miss the fainter meteors, and your results will be inaccurate.

Meteorites, solid objects which can survive the complete drop to the ground, are different from shooting-star meteors. They are not connected with comets, and are more nearly related to the asteroids; there may be no difference between a large meteorite and a small asteroid. Most museums have collections of them, but the largest known meteorite is still lying where it fell in prehistoric times at Hoba West, near Grootfontein in South Africa. I doubt whether anyone will run away with it, since it weighs at least 60 tons. The largest meteorite in captivity, so to speak, is the 36-ton Greenland monster discovered by the Arctic explorer Robert Peary, and brought laboriously to the United States, where it is now on view at the Hayden Planetarium in New York.

Meteorites are of two main types, stones (aerolites) and irons (siderites). Really large objects may produce craters. The best-known example is Meteor Crater in Arizona (really it should be called Meteor*ite* Crater), which is nearly a mile wide; it was formed about 50,000 years ago, and is a tourist attraction, so that if you are ever in that part of the world I urge you to go and see it. Wolf Creek, in one of the wilder parts of Australia, is another meteorite crater. The Vredefort Ring, near Pretoria in South Africa, is often listed as being due to a major fall, but geologists who have made prolonged studies of it are unanimous in regarding it as volcanic. The Sacred Stone in the holy city of Mecca is, however, undoubtedly a meteorite.

The last large meteorites seen to pass over Britain were those of Christmas Eve 1965, when a meteorite came down near Barwell in Leicestershire, and in 1969, when another meteorite passed over England, Wales and Ulster before ending its career in the sea. The Barwell Meteorite was widely observed during its flight; pieces of it showered down in and near the village. One fragment was later found inside a house, nestling coyly in a vase of artificial flowers; evidently it had made its entry via an open window. On 5 May 1992 a very small meteorite fell at Glatton in Cambridgeshire.

Nobody has ever been killed by a tumbling meteorite, so far as we know; the only casualty has been an Egyptian dog, in 1911, which was in the wrong place at the wrong moment. Of course, meteorites have been known since very early times, though it was only in the eighteenth century that scientists decided that they come from the sky.

There are also tektites – small, glassy objects found in restricted areas of the world, which are decidedly mysterious, but which are now thought to be of terrestrial origin.

It is difficult for the non-expert to tell a meteorite from a non-meteorite. Some years ago I devoted a television programme to them, with dire results. Over the next few weeks I was inundated with pieces of material of all shapes and sizes; they arrived by every post, with letters asking in various ways, 'Is this a meteorite? And if so, what is it worth?' Not one 'meteorite' was genuine, but I accumulated such bizarre specimens as a piece of ship's anchor, several slabs of cork matting, a couple of cannonballs and a very ancient Bath bun. By the time I had returned them to their owners, my enthusiasm for meteorites had been somewhat dampened. So if you think that you have found one, notify me by all means, but please don't send it through the post!

37

Glows in the sky

On the evening of 26 January 1938, when the world was still in a state of uneasy peace, the daily newspapers in England were full of an event which was quite unconnected with politics. 'Weirdest Storm Over London', read one headline. 'Sky Glows Fiery Red', went another. There followed a long description of what many people had taken to be the glare of a great fire, but which was in fact a particularly brilliant display of the Northern Lights. For once, Hitler and Mussolini were mercifully relegated to the back pages.

I watched that display from my Sussex home, and even now it remains vividly in my mind, though from Norway and northern Canada I have seen more spectacular aurorae since. It is true that the sky turned red, but there were arches, streamers and shining curtains as well, shifting and changing all the time. Nobody who saw it can have failed to be impressed.

Aurorae have been known for centuries. The Chinese were familiar with them, and there is an interesting note in an old album which seems to refer to an aurora seen around 2600 BC: 'The mother of the Yellow Emperor Shuan-Yuan, Fu-Fao, saw a big lightning circulating round the Su star of Bei-Dou, with the light shining all over the field. She then became pregnant.' The star concerned was certainly Dubhe in Ursa Major, though whether the aurora had anything to do with Mrs Fu-Pao's coming happy event seems more dubious! In AD 555 Matthew of Westminster wrote of 'lances seen in the air', perhaps the first British reference to aurorae.

The name 'aurora borealis' dates from 1621, when it was coined by the French astronomer Pierre Gassendi. In the south we have an aurora australis, or Southern Light. The aurora australis appears in the Maori legends of New Zealand, and is said to be due to a great fire lit by the Maori ancestors whose canoes had drifted as far south as the Antarctic Ocean.

Over the centuries, the aurora has had many names. In Ancient Greece it was sometimes termed 'blood rain'; the Germans called it 'heaven light'. The Eskimo name was 'sky dwellers' (selamuit), and in the Shetland Isles, off Scotland, aurorae were known as the Merry Dancers. There are fewer names for aurora australis, because it is not seen so widely over inhabited lands.

But what exactly are aurorae?

Though they occur in the Earth's upper air, they are certainly the concern of the astronomer, because they are due to particles sent out by the Sun. I have referred earlier to the solar wind, which has such a marked effect upon the tails of comets. The particles are electrically charged, so that when they encounter the Earth they spiral down towards the magnetic poles – or, to be more precise, to two oval-shaped regions, one 23 degrees from the north pole and the other 23 degrees from the south pole. The mechanism is not so straightforward as used to be thought, and the zones of radiation round the Earth called the Van Allen zones (identified in 1958 from the first American artificial satellite, Explorer 1) are involved in the process, but the Sun is wholly responsible.

This means that aurorae are best seen from high latitudes. The percentage of nights in which aurorae maybe expected ranges from 100 in Barrow (Alaska) and Churchill (Canada) to 60 in Winnipeg, 10 in Montreal, 4 in New York, below 2 in San Francisco and below 1 in Houston; 90 at Tromsø in Norway, 10 in Oslo, 8 in Edinburgh, below 3 in London, and almost never in Rome or Athens. In the south, Melbourne may expect about three 'aurora nights' annually, but Sydney and Auckland only one, and aurorae are seldom visible from Cape Town. So far as I know, the aurora of 25 September 1909 was the most widespread on record. It was definitely visible from Djakarta in Java, latitude 6° S, and also, it seems, from Singapore, latitude 1° N.

Apparently there is no difference between the aurorae of the two poles. Our records of the aurora australis are much less complete, for obvious reasons, but now that there are permanent stations in Antarctica there is continuous monitoring. Aurorae are commonest near the peak of the 11-year sunspot cycle, as for example around 2000–1.

There are several distinct types of aurorae. First there is the general horizon glow, which is none too easy to identify; it is easy to be deceived by the glow of a distant town. If the light extends upward from the horizon to form an arc, there can be no doubt about its nature. If the display is a major one, the arc may then send up rays at right-angles to its length, while the arc itself folds to form an irregular band. When the rays are really long, the band gives the impression of what is termed drapery, and the rays may converge at a point at the zenith, forming an auroral corona (nothing to do with the Sun's corona). Waves of light may surge up from the horizon; this so-called flaming aurora is glorious. There may also be a diffuse sky-glow or veil, as well as diffuse isolated patches.

For observing aurorae, telescopes are quite useless, and binoculars nearly so. The best instrument is the human eye, and nothing is needed except a rough measuring device. A foot-rule held at arm's length will give an angle of one degree to the half-inch, and this is usually good enough, though an elaborate quadrant can be made by anyone with sufficient practical skill. It is also useful to remember that the distance between the Pointers in Ursa Major is 5 degrees, while Polaris to Beta Cassiopeiae is 30 degrees.

Naked-eye observations of aurorae are valuable, and negative evidence is also important. If you make a check and find that no aurora is visible, note the time and the condition of the sky. (Since aurorae range in height between 60 miles and over 500 miles they are, of course, well above any normal clouds.) When a display is seen, the form and position should be written down; the position angle is given in degrees, starting at 0 for north and working round to east (90), south (180) and west (270) back to north. Together with the altitude, this will fix the position of the main aurora, but it is often sufficient to do no more than mark it against the background constellations.

Peculiar crackling sounds have been reported from time to time during auroral displays. There is no obvious reason why an aurora should be audible, and I admit to being sceptical, but it is worth listening out. (I have also had reports about pungent odours during auroral activity. I am even more sceptical about smelly aurorae than I am about noisy aurorae, but it is always dangerous to be dogmatic!)

Aurorae are splendid photographic subjects. The other important sky-glow, the Zodiacal Light, is much dimmer and is purely astronomical, because it is due to small particles spread along the main plane of the Solar System and lit up by the Sun. The Light is confined to regions near the ecliptic. Near the horizon, the width of the main cone may be as much as 25 degrees, but it tapers quickly and becomes narrow as it stretches upward into the sky. Its name is appropriate, since it lies in the Zodiacal region.

The Light is not easy to observe from Britain or the northern United States. It is not affected by any magnetic force; it is never visible until the sky is dark, and neither can it remain visible late at night, for by then the Sun has dropped well below the horizon and the Zodiacal Light has also set. You have to look for it at the most favourable moments; and to make things worse, moonlight will drown it completely.

The Light is most often on view when the ecliptic is nearest to the perpendicular to the horizon. Under really clear skies, as you will find in countries nearer the equator. the Light is said to become brighter than the Milky Way.

To observe the Zodiacal Light, all you need is a star-map together with a one-foot rule. Record the width of the base in degrees (remembering that at arm's length one degree is equal to half an inch on the ruler); estimate the altitude of the top of the cone in degrees above the horizon; then plot the cone itself on a star-map as accurately as possible, and give the intensity of the Light as compared with different parts of the Milky Way. It is important to gauge the transparency of the sky, which is best done by noting the faintest stars visible and then checking on their magnitudes.

Some observers have found that the main cone has definite structure, with the inner core surrounded to either side by fainter flanks. I have never seen this myself, and neither have I seen the slightly pinkish hue of the inner region which has been described on various occasions, but observers with keener sight than mine should be able to do better.

If you are going to make useful records of the Zodiacal Light, you must have a good working knowledge of the constellations, so that the position of the dim, ghostly cone can be checked. It is useless to say that the apex 'stretches up to a bright star with a fainter one beside it'. If you can say 'The apex lies midway between Zeta and Beta Tauri', the position can be plotted accurately.

There are two other glows of the same basic nature as the Zodiacal Light, though they are much more elusive. One is the Zodiacal Band, a parallel-sided beam of extremely faint, diffused light stretching across the sky and making up what is in effect an extension of the main cone. It is nothing like so bright as the Milky Way, and cannot be seen at all if it lies near the Milky Way area. Even the slightest trace of mist or moonlight will conceal it.

I have never seen the Zodiacal Band properly, but I have glimpsed the Gegenschein, or counterglow, which was first described in detail by the Danish astronomer Theodor Brorsen over a century ago. It also is due to thinly spread material near the main plane of the Solar System. It lies exactly opposite to the Sun in the sky, and on occasion it may cover an area as large as that of the Square of Pegasus. My one good observation of it was made from an airfield in 1941, when England was blacked out and the sky was really dark. I have often looked for it since, but with no result. Those who live far away from towns should do better, but absolute darkness and clarity are essential. The Gegenschein is extremely difficult to photograph..

Remember, too, that the human eye is slow to adapt. There is no point in walking out from a brilliantly lit room and expecting to see faint glows (or, for that matter, faint stars). You will have to stay in the dark for some time before your eye becomes fully adapted. Beware, also, of the familiar pocket torch. As soon as you switch on a light to record your notes, the process of adapting to the dark will have to be started all over again. If you must use a light, fit your torch with a dim red bulb. My own practice is to record my notes on a portable tape-recorder, so that they can be transcribed later.

I must mention the Kordylewski Clouds, first reported by the Polish astronomer K. Kordylewski in 1961. He believed them to be due to collections of interplanetary débris moving in the same orbit as the Moon, either 60 degrees ahead or 60 degrees behind. It has been claimed that the Clouds have been seen with the naked eye as excessively faint patches of light, but their existence has yet to be confirmed. I have often looked for them, but with a signal lack of success.

There are various other sky-glows. Anything inside the atmosphere is really the province of the meteorologist rather than the astronomer (apart from the aurora, which has a purely solar origin). Still, it may be useful to say a little about them.

Rainbows are caused by water-droplets in the air, which split up the Sun's light; often there are two bows, a primary and a secondary. In the primary, red is seen at the outer part of the band, with violet at the inner edge and the other colours in between; in the fainter secondary, red is on the inside. Obviously a rainbow must always lie opposite to the Sun, and it is always a sign of showery weather.

It is not so generally known that the Moon also can cause rainbows, though without detectable colour. I once saw an excellent lunar rainbow when I was flying over north Scotland during the early part of World War II, but I was unable to pay much attention to it, for the excellent reason that I was navigating the aircraft and had no particular desire to find myself over Cologne instead of Caithness. I have seen two since, but they are certainly rare.

A Sun Pillar is due to the reflection of the Sun's rays from the vertical sides of a column of ice crystals in the atmosphere. It looks like a pillar of red or white light extending vertically above and below the Sun, sometimes crossed by a similar horizontal bar making up what has been called a Heavenly Cross. Moon Pillars are also known.

Haloes are produced when the Sun or Moon shines through a very thin layer of cloud of the type known as cirrostratus, which forms a veil over the sky and is made up of ice crystals at over 20,000 feet above ground level. The cloud itself is more or less transparent, and is not easy to detect, but it is responsible for the haloes, which appear as faint rings surrounding the Sun or Moon. Colours are not usually conspicuous, but there may be a red tint on the inner edge of the halo and a yellow cast on the outer part. Cirrostratus cloud is often the forerunner of rain, so that it is quite correct to regard a halo as a sign of bad weather on the way.

A corona (not to be confused with the Sun's corona or the Zodiacal corona) is associated with lower clouds, altostratus, and is due to the diffraction of light rays by water particles inside the cloud. Coronae are smaller than haloes, and are not necessarily indicative of approaching rain. Lunar coronae may be spectacular at times.

When the weather is very cold and ice crystals form in the lower part of the atmosphere, refraction effects produce false images of the Sun, often joined by luminous rays. These are known as parhelia, or mock-suns, and are best seen from the polar zones. Mock-moons are similar, though much less striking.

Finally, them are blue moons and blue suns, which do occur occasionally when the upper air contains an unusual amount of dust or volcanic ash. I have seen several, the most impressive being that of 26 September 1950, when from Sussex the Moon appeared a lovely, shimmering blue. The cause was dust in the upper atmosphere produced by gigantic forest fires in Canada. The tremendous volcanic eruption of Krakatoa, in 1883, is said to have produced a host of coloured moons and suns, together with peculiar sunsets; there were also noticeable effects after the recent eruption of Mount St Helens in Washington.

With the true astronomical glows, city-dwellers are at a hopeless disadvantage, but anyone who can drive out into the country should be able to see them. And if you are lucky enough to catch sight of the dim, tapering cone of the Zodiacal Light, stretching upward from the horizon into the blackness of the sky, you will feel that your journey has been well worth while. The Light may not be spectacular, but it has a faint, eerie beauty all its own.

I hope that what I have said will be enough to show that there is always plenty to be seen in the sky. Astronomy is open to everyone, and if you want to take a real interest you can do so without the need for spending a large sum of money on a powerful telescope.

Appendices

I THE CONSTELLATIONS

Latin name	*English name*	*First-magnitude star(s)*
Andromeda	Andromeda	
Antlia	The Air-Pump	
Apus	The Bee	
Aquarius	The Water-Bearer	
Aquila	The Eagle	Altair
Ara	The Altar	
Aries	The Ram	
Auriga	The Charioteer	Capella
Boötes	The Herdsman	Arcturus
Caelum	The Sculptor's Tools	
Camelopardalis	The Giraffe	
Cancer	The Crab	
Canes Venatici	The Hunting Dogs	
Canis Major	The Great Dog	Sirius
Canis Minor	The Little Dog	Procyon
Capricornus	The Sea-Goat	
Carina	The Keel	Canopus
Cassiopeia	Cassiopeia	
Centaurus	The Centaur	Alpha Centauri, Agena
Cepheus	Cepheus	
Cetus	The Whale	
Chamaeleon	The Chameleon	
Circinus	The Compasses	
Columba	The Dove	
Coma Berenices	Berenice's Hair	
Corona Australis	The Southern Crown	
Corona Borealis	The Northern Crown	
Corvus	The Crow	
Crater	The Cup	
Crux Australis	The Southern Cross	Acrux, Beta Crucis
Cygnus	The Swan	Deneb
Delphinus	The Dolphin	

Latin name	*English name*	*First-magnitude star(s)*
Dorado	The Swordfish	
Draco	The Dragon	
Equuleus	The Little Horse	
Eridanus	The River	Achernar
Fornax	The Furnace	
Gemini	The Twins	Pollux
Grus	The Crane	
Hercules	Hercules	
Horologium	The Clock	
Hydra	The Watersnake	
Hydrus	The Little Snake	
Indus	The Indian	
Lacerta	The Lizard	
Leo	The Lion	Regulus
Leo Minor	The Little Lion	
Lepus	The Hare	
Libra	The Balance	
Lupus	The Wolf	
Lynx	The Lynx	
Lyra	The Lyre	Vega
Mensa	The Table	
Microscopium	The Microscope	
Monoceros	The Unicorn	
Musca Australis	The Southern Fly	
Norma	The Rule	
Octans	The Octant	
Ophiuchus	The Serpent-Bearer	
Orion	Orion	Rigel, Betelgeux
Pavo	The Peacock	
Pegasus	The Flying Horse	
Perseus	Perseus	
Phoenix	The Phoenix	
Pictor	The Painter	
Pisces	The Fishes	
Piscis Austrinus	The Southern Fish	Fomalhaut
Puppis	The Poop	
Pyxis	The Compass	
Reticulum	The Net	
Sagitta	The Arrow	
Sagittarius	The Archer	
Scorpius	The Scorpion	
Sculptor	The Sculptor	
Scutum	The Shield	
Serpens	The Serpent	
Sextans	The Sextant	
Taurus	The Bull	Aldebaran
Telescopium	The Telescope	

Latin name	*English name*	*First-magnitude star(s)*
Triangulum	The Triangle	
Triangulum Australe	The Southern Triangle	
Tucana	The Toucan	
Ursa Major	The Great Bear	
Ursa Minor	The Little Bear	
Vela	The Sails	
Virgo	The Virgin	Spica
Volans	The Flying Fish	
Vulpecula	The Fox	

II STAR NAMES

Generally speaking, the proper names of the stars are used only for those of the first magnitude and a few other special cases, such as Mizar, Algol and Mira. However, some people are still fond of them, and it may be useful to give a list here of all stars down to the third magnitude which are named, plus a few others.

ANDROMEDA	Alpha:	Alpheratz
	Beta:	Mirach
	Gamma:	Almaak
AQUARIUS	Alpha:	Sadalmelik
	Beta:	Sadalsuud
AQUILA	Alpha:	Altair
	Beta:	Alshain
	Gamma:	Tarazed
	Zeta:	Dheneb
ARIES	Alpha:	Hamal
	Beta:	Sheratan
	Gamma:	Mesartim
AURIGA	Alpha:	Capella
	Beta:	Menkarlina
	Iota:	Hassaleh
BOÖTES	Alpha:	Arcturus
	Epsilon:	Izar
CANES VENATICI	Alpha:	Cor Caroli
CANIS MAJOR	Alpha:	Sirius
	Beta:	Mirzam
	Delta:	Wezea
	Epsilon:	Adhara
	Eta:	Aludra
CANIS MINOR	Alpha:	Procyon
CAPRICORNUS	Delta:	Deneb al Giedi
CARINA	Alpha:	Canopus
	Beta:	Miaplacidus

	Epsilon:	Avior
	Iota:	Tureis
CASSIOPEIA	Alpha:	Shedir
	Beta:	Chaph
CENTAURUS	Beta:	Agena
	Theta:	Haratan
CEPHEUS	Alpha:	Alderamin
CETUS	Alpha:	Menkar
	Beta:	Diphda
	Omicron:	Mira
COLUMBA	Alpha:	Phakt
CORONA BOREALIS	Alpha:	Alphekka
CORVUS	Beta:	Kraz
	Gamma:	Minkar
	Delta:	Algorel
CRUX AUSTRALIS	Alpha:	Acrux
CYGNUS	Alpha:	Deneb
	Beta:	Albireo
	Gamma:	Sadr
	Epsilon:	Gienah
DRACO	Alpha:	Thuhan
	Beta:	Alwaid
	Gamma:	Eltamin
ERIDANUS	Alpha:	Achernar
	Beta:	Kursa
	Gamma:	Zaurak
	Theta:	Acamar
GEMINI	Alpha:	Castor
	Beta:	Pollux
	Gamma:	Alhena
	Delta:	Wasat
	Epsilon:	Mebsuta
	Zeta:	Mekbuda
	Eta:	Propus
GRUS	Alpha:	Alnair
	Beta:	Al Dhanab
HERCULES	Alpha:	Rasalgethi
	Beta:	Kornephoros
	Zeta:	Rutilicus
HYDRA	Alpha:	Alphard
LEO	Alpha:	Regulus
	Beta:	Denebola
	Gamma:	Algieba
	Theta:	Chort
LEPUS	Alpha:	Arneb
	Beta:	Nihal
LIBRA	Alpha:	Zubenelgenubi
	Beta:	Zubenelchemale

	Sigma:	Zubenalgubi
LYRA	Alpha:	Vega
	Beta:	Sheliak
	Gamma:	Sulaphat
OPHIUCHUS	Alpha:	Rasalhague
	Beta:	Cheleb
	Zeta:	Han
	Eta:	Sahik
ORION	Alpha:	Betelgeux
	Beta:	Rigel
	Gamma:	Bellatrix
	Delta:	Mintaka
	Epsilon:	Alnilam
	Zeta:	Alnitak
	Iota:	Hatysa
	Kappa:	Saiph
PEGASUS	Alpha:	Markab
	Beta:	Skat
	Gamma:	Algenib
	Epsilon:	Enif
PERSEUS	Alpha:	Mirphak
	Beta:	Algol
PHOENIX	Alpha:	Ankaa
PISCIS AUSTRALIS	Alpha:	Fomalhaut
PUPPIS	Zeta:	Suhail Hadar
	Rho:	Turais
SAGITTARIUS	Gamma:	Alnasr
	Delta:	Kaus Meridionalis
	Epsilon:	Kaus Australis
	Zeta:	Ascella
	Lambda:	Kaus Borealis
	Sigma:	Nunki
SCORPIUS	Alpha:	Antares
	Beta:	Graffias
	Epsilon:	Wei
	Theta:	Sargas
	Kappa:	Girtab
	Upsilon:	Lesath
	Sigma:	Alniyat
SERPENS	Alpha:	Unukalhai
TAURUS	Alpha:	Aldebaran
	Beta:	Al Nath
	Epsilon:	Ain
	Zeta:	Alheka
URSA MAJOR	Alpha:	Dubhe
	Beta:	Merak
	Gamma:	Phad
	Delta:	Megrez

	Epsilon:	Alioth
	Zeta:	Mizar
	Eta:	Alkaid
	80:	Alcor
URSA MINOR	Alpha:	Polaris
	Beta:	Kocab
VELA	Gamma:	Regor
	Kappa:	Markeb
	Lambda:	Al Suhail al Wazn
VIRGO	Alpha:	Spica
	Gamma:	Arich

III MONTHLY STAR MAP REFERENCE TABLE

The table indicates the appropriate star map for any time or month in the year. The left-hand column gives the month of observation, while the horizontal column represents the time of observation to the nearest 2 hours. By simply reading across and downward, the required star map can be found: thus the map for a midnight observation in February will be number 3.

NORTHERN MAPS (1 to 12)

G.M.T.	1800 6 pm	2000 8 pm	2200 10 pm	0000 midnight	0200 2 am	0400 4 am	0600 6 am
January	11	12	1	2	3	4	5
February	12	1	2	3	4	5	6
March	1	2	3	4	5	6	7
April		3	4	5	6	7	
May		4	5	6	7	8	
June			6	7	8	9	
July		6	7	8	9	10	
August		7	8	9	10	11	
September		8	9	10	11	12	
October	8	9	10	11	12	1	2
November	9	10	11	12	1	2	3
December	10	11	12	1	2	3	4

SOUTHERN MAPS (13 to 24)

	6 pm	8 pm	10 pm	midnight	2 am	4 am	6 am
January	23	24	13	14	15	16	17
February	24	13	14	15	16	17	18
March	13	14	15	16	17	18	19
April	14	15	16	17	18	19	20

May	15	16	17	18	19	20	21
June	16	17	18	19	20	21	22
July	17	18	19	20	21	22	23
August	18	19	20	21	22	23	24
September	19	20	21	22	23	24	13
October	20	21	22	23	24	13	14
November	21	22	23	24	13	14	15
December	22	23	24	13	14	15	16

IV ARTIFICIAL SATELLITES

The first artificial satellite, or man-made moon, was Sputnik 1, launched by the Russians on 4 October 1957. Since then there have been many others, some of them clearly visible with the naked eye as slowly moving, starlike points. They are much too slow to be confused with meteors, but they can be mistaken for stars if their movements are gradual enough,

In the early years of the Space Age, little was known about the density of the uppermost part of the atmosphere, in which some of the satellites moved. Observations of the satellites were therefore most important, and Project Moonwatch was set up by NASA. It was mainly amateur, but it was of tremendous value and operated for many years before it was finally 'stood down'. I had many letters from people who reported that they had seen a satellite suddenly vanish. The reason was that it had passed into the Earth's shadow, so that its supply of direct sunlight was cut off. Also, many observers were puzzled by apparently irregular wobblings of the satellites. The usual cause was the rotation of a non-spherical satellite.

Observations of satellites may be less useful now than they used to be, but they are still worth making. It is surprisingly seldom that a satellite passes directly in front of a star visible with the naked eye, and the method is to use a stop-watch, starting it at the moment when the satellite either passes close to a known star or else tracks between two stars which are suitably placed. Then, as quickly as possible, check the time either by radio or, if need be, by dialling the Speaking Clock on the telephone and stop the watch. Suppose that you stop the watch at 20h 11m 51.7s, and find that it reads 1m 20.0s. This means that the satellite fix was obtained 1m 20s earlier – that is to say, at 20h 10m 31.7s. What you are doing is using the stop-watch to time accurately the interval from the instant the event occurs to the instant you obtain a precise time over the telephone. By subtracting the stop-watch interval from telephone time, you obtain the exact time of your satellite observation. This procedure can be used to time all kinds of astronomical events, not just satellites.

It is vital to check the stop-watch frequently, to make sure that it is keeping good time. Also, be careful about your comparison stars. It is not helpful to record a satellite as passing by, say, Zeta Aquilae when it is actually passing near Gamma Draconis.

V ASTRONOMICAL SOCIETIES

If you have been interested in what you have read in this book, I strongly urge you to join an astronomical society. Most major towns and cities in Britain and America have their own local clubs. Those in Britain are listed in the annual *Yearbook of Astronomy*. The main British observational society is the British Astronomical Association, which was formed in 1890 and has a record second to none. No actual qualifications are needed for membership. Meetings are held regularly in London, and nowadays in other centres also; there are many publications, including a bi-monthly journal, and there are special observing sections. Full information can be obtained from the Assistant Secretary at Burlington House, Piccadilly, London W.1.

In the United States, the extensive astronomical activities are covered in *Sky and Telescope* and *Astronomy* magazines available at any major library. Also, virtually every larger city, as well as many smaller ones, has a planetarium which is usually the centre of astronomical activities. The planetarium staff will be pleased to supply details about any local astronomical society, whether or not it meets at the planetarium.

Most other countries have active national astronomical societies.

Index

Page references to constellations are given in bold type.